W9-AMQ-393

CIVILIZED LIFE IN THE UNIVERSE

CIVILIZED LIFE IN THE UNIVERSE

SCIENTISTS ON INTELLIGENT EXTRATERRESTRIALS

GEORGE BASALLA

2006

OXFORD
UNIVERSITY PRESS

Oxford University Press, Inc., publishes works that further
Oxford University's objective of excellence
in research, scholarship, and education.

Oxford New York
Auckland Cape Town Dar es Salaam Hong Kong Karachi
Kuala Lumpur Madrid Melbourne Mexico City Nairobi
New Delhi Shanghai Taipei Toronto

With offices in
Argentina Austria Brazil Chile Czech Republic France Greece
Guatemala Hungary Italy Japan Poland Portugal Singapore
South Korea Switzerland Thailand Turkey Ukraine Vietnam

Published by Oxford University Press, Inc.
198 Madison Avenue, New York, New York 10016

www.oup.com

Oxford is a registered trademark of Oxford University Press

Library of Congress Cataloging-in-Publication Data
Basalla, George.
Civilized life in the universe : scientists on intelligent extraterrestrials / George Basalla.
p. cm.
Includes bibliographical references and index.
ISBN-13 978-0-19-517181-5
ISBN 0-19-517181-0
1. Life on other planets—Research. 2. Outer space—Exploration—Public opinion.
I. Title.
QB54.B37 2005
576.8'39—dc22 2005040677

1 3 5 7 9 8 6 4 2

Printed in the United States of America
on acid-free paper

For Carolyn

ACKNOWLEDGMENTS

I gratefully acknowledge the research done by two first-rate historians of extraterrestrial life: Michael J. Crowe and Steven J. Dick. They combine meticulous research with judicious interpretation of documents to produce outstanding works of scholarship. I have followed their lead again and again even though our emphases, interpretations, and conclusions differ at times.

Many other scientists, historians, and writers, too numerous to mention individually, have influenced my interpretation of extraterrestrial civilizations. The specific contributions of these authorities are recorded in the bibliographies, keyed to each section of a chapter, presented at the end of the text. I am indebted to the authors cited for my understanding of this complex and controversial topic.

I owe special thanks to the staff of the Morris Library at the University of Delaware, where I did all of my research for this book. I have drawn upon the patience, knowledge, and professional skills of librarians ably assembled and directed by the May Morris Director of Libraries, Susan Brynteson. The Morris Library provided me with the physical facilities, information sources, professional attention, and intellectual stimulus to carry out my project.

I acknowledge the source of each illustration in the caption accompanying it. However, I must single out the exemplary help I received from the Lowell Observatory, Flagstaff, Arizona. Lowell personnel, especially librarian Antoinette Beiser, made a special effort to help me obtain images of Mars and Martian canals.

Finally, I am indebted to my editor, Clifford Mills, along with John Rauschenberg, of the Oxford University Press. They, along with independent editor Lyman Lyons, helped to shape my conception of the book and make the text more accessible to readers. Cliff Mills smoothed over many of the problems I brought to him with sympathy and continuous encouragement. I thank him for his friendship and guidance.

Every effort has been made to locate the holders of rights to the illustrations that appear in this book; we regret if any have been inadvertently overlooked.

CONTENTS

INTRODUCTION

In 1980 astronomer Carl Sagan presented a popular television series (*Cosmos*) in which he guided viewers on a spectacular tour of the universe. The series covered 15 billion years of cosmic evolution, including the birth of the stars, the origins of our solar system, the emergence of intelligent life on Earth and on worlds in outer space, the history of humanity, and the development of space science and technology.

Near the end of *Cosmos*, Sagan estimated the number of advanced technological civilizations thriving in the Milky Way Galaxy. He said there were millions of civilizations scattered throughout our Galaxy and that interstellar space was filled with radio messages sent by extraterrestrial transmitters. The messages constitute an *Encyclopedia Galactica,* the knowledge and wisdom gathered by millions of civilizations over millions of years of Galactic history.

According to Sagan, the *Encyclopedia Galactica* was ours to discover. By properly orienting large numbers of radio telescopes to capture signals coming from deep space, humans could gain access to the *Encyclopedia.* Here they would find solutions to many problems troubling modern society: war, environmental pollution, natural resource depletion, overpopulation, energy shortages, and so on.

This book explains how prominent scientists like Carl Sagan came to believe in the existence of superior alien civilizations and the importance of contacting them. It draws upon the works of respected astronomers, physicists, chemists, and biologists who speculated about extraterrestrial civilizations for more than 400 years. Their conjectures appear in articles and books written to advance scientific knowledge and educate the general public.

Civilization implies the existence of intelligent creatures who create complex social and cultural institutions and cultivate science and technology. Because I concentrate on civilized beings, I have little to say about the origin of primitive life on Earth or the possible existence of microscopic life elsewhere in the universe. And, since I emphasize scientific accounts of extraterrestrial civilizations, I make limited use of the literature of science fiction, UFO visitation, and alien abduction.

There is no fixed boundary line separating scientific perceptions of extraterrestrial civilizations from popular treatments of the subject. Important features of the scientific depiction of advanced life on other worlds appear in popular culture. These shared visions originate in ancient streams of thought that nourished both scientific and popular ideas about intelligent extraterrestrial beings, and continue to influence it today.

The pursuit of alien beings began in antiquity when Greek philosophers first considered the nature of the universe and its possible inhabitants. Medieval thinkers, working within the Christian tradition, continued to speculate about these matters. Pagan and Christian thinkers agreed that the Earth was located at the center of the universe

Centuries later, at the time of Europe's Renaissance, Nicolaus Copernicus established the Sun-centered universe. The Copernican astronomical revolution, and the invention of the optical telescope at the beginning of the seventeenth century, stimulated the belief that humans might find proof of intelligent creatures on the Moon, or on a planet in the solar system. Astronomers carefully studied the lunar landscape with their instruments but finally concluded that the Moon was empty of life.

As the magnifying power of telescopes increased, and scientific knowledge of astronomical bodies developed, astronomers turned their attention to close-by planets, especially Mars. The telescopic observation of Mars yielded spectacular results in the late nineteenth century. Several respected astronomers, notably Giovanni Schiaparelli and Percival Lowell, claimed to see a complex system of irrigation canals on the Martian landscape. The canals were hailed as proof of advanced Martian engineering.

Many astronomers, however, failed to observe the canals, or dismissed them as the result of physical, physiological, and psychological factors that affected what observers saw on a planet located at least 35 million miles from Earth. The dispute over the canals and Martian life continued until the mid-1960s when spacecraft sent back images of the planet that contained no signs of canals.

The search for intelligent life in the universe was enlarged by the invention of the radio telescope in the mid-twentieth century. An instrument designed to collect radio emissions from celestial bodies can also detect radio messages sent by the intelligent creatures living on planets beyond the range of optical telescopes.

Since its invention, the radio telescope has been the preferred instrument for scientists who search for extraterrestrial intelligence (SETI).

SETI scientists, who include Frank Drake, Philip Morrison, and Carl Sagan, successfully convinced America's space agency, the National Aeronautics and Space Administration (NASA), to support their venture. NASA funding introduced SETI into the political arena and the scrutiny of members of Congress who control NASA's budget. National recognition of SETI also exposed it to critics who were skeptical of claims that advanced extraterrestrial civilizations were transmitting important messages to Earth.

Astronomers use telescopes, optical or radio, to gather evidence for the existence of intelligent life in the universe. The process of data collection has generated disputes among scientists, but the interpretation of existing data, or reasons for the lack of relevant data, has been the main source of controversy for centuries. Intelligent alien beings remain as elusive, and controversial, in the twenty-first century as they were in ancient times when Greek thinkers first wondered about their existence.

CIVILIZED LIFE IN THE UNIVERSE

CHAPTER ONE

Trio of Ideas

> What I am more concerned with is the extent to which the modern search for aliens is, at rock-bottom, part of an ancient religious quest.
>
> —Paul Davies, *Are We Alone?* 1995

The Ideas

Scientific perceptions of advanced extraterrestrial life are based upon a trio of ideas that first appeared in the religious and philosophical thought of antiquity and the Middle Ages. The first idea is that the universe is very large, if not infinite in extent. The second, that we are not alone in the universe, there are other inhabited worlds somewhere in the vastness of space. The third, that there is an essential difference between the superior beings of the celestial world and the inferior ones who live on Earth.

These three ideas are relevant to the work of scientists today. Modern cosmologists have determined that the universe is expanding at an increasing rate and is unlikely to slow down and collapse on itself in a final Big Crunch. Within our immense universe, astronomers have recently identified more than 100 extrasolar planets. An extrasolar planet is one that orbits a star located far beyond our solar system. Some scientists believe that extrasolar planets are inhabited by creatures with a level of intelligence and civilization that surpasses the intellect and civilized life of humans. Astronomers, however, have just begun their investigations and have found no evidence of extraterrestrial civilizations.

Any examination of extraterrestrial civilizations must begin with the debt modern science owes to the trio of ideas that shaped our ways of thinking about the universe and its inhabitants. These key assumptions, which appear so often in the modern search for extraterrestrial intelligence, arose in much earlier times and within different contexts.

The Infinitization of the Universe

The ancient Greek atomists were among the first to introduce the idea of an infinite universe. In the fifth century B.C., they claimed that tiny bits of matter (atoms) moved randomly in infinite empty space. Because an infinite number of atoms collide an infinite number of times in an infinite void, an infinite number of universes exist. Each universe has its own sun, planets, stars, and life forms.

A century later the influential philosopher Aristotle (384–322 B.C.) rejected the atomists' infinite void and their many universes. In its place, he put a single finite universe with the Earth located at its exact center. The planets, Sun, and stars all circle the motionless Earth.

The stars marked the outer limits of Aristotle's geocentric (Earth-centered) universe, and he refused to consider the existence of space beyond the stellar boundaries. There are no voids, vacuums, or empty spaces in Aristotle's universe because the region between the Moon and the stars is filled with a solid, transparent, crystalline material.

Aristotle's view of the universe explained all known astronomical phenomena and satisfied the ordinary observer's feeling that the Earth was at the center of things. It lasted for nearly two thousand years and inspired some of the greatest scientific, philosophical, theological, and literary minds in Western civilization.

By the fourteenth century, however, critics argued that Aristotle was wrong to place limits on God by confining Him to a finite universe. God extends Himself, they said, filling *infinite* space with His *immensity*. Influenced by their conception of an infinite God, philosophers and theologians in the Middle Ages accepted a universe that was infinite in extent. The identification of God with infinite space, sometimes called the divinization of space, lasted well into the seventeenth century.

Some Renaissance astronomers and philosophers were not satisfied with the medieval understanding of the cosmos. They insisted that the universe was

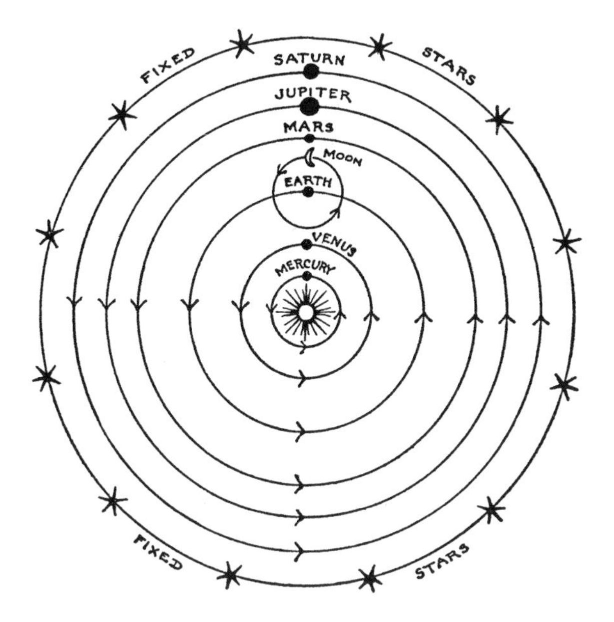

FIG. 1.1. Sun-centered, finite universe proposed by Copernicus in 1543. (Angus Armitage, *Sun, Stand Thou Still.* New York: Henry Schuman, 1947.)

more than a theological construct. By interpreting astronomical observations mathematically, they argued, it was possible to obtain a true picture of physical reality. The crucial figure in this intellectual revolution was the Polish astronomer and Church administrator Nicolaus Copernicus (1473–1543). He proposed a heliocentric (Sun-centered) model of the universe. It featured a stationary Sun at the center of a system of orbiting planets that included the Earth (Fig. 1.1). The Copernican universe remained finite, but it was substantially larger than the old geocentric model made popular by Aristotle.

The infinitization of the universe grew out of the fundamental changes Copernicus made in the arrangement of the Sun, Earth, and planets and in the motions of the Earth. By the second half of the sixteenth century, followers of Copernicus claimed that the universe extended to infinity. The first printed illustration of an infinite universe dates to 1576, just thirty-three years after Copernicus published his theory of a heliocentric universe (Fig. 1.2).

Most astronomers, if not the general public, soon accepted the infinite nature of the universe. In the seventeenth century, Sir Isaac Newton made a static infinite universe an integral part of his new physics of moving bodies on Earth and in the heavens. Despite the work of generations of physicists and astronomers who succeeded Newton, the precise nature of the universe remains an unresolved problem for modern science.

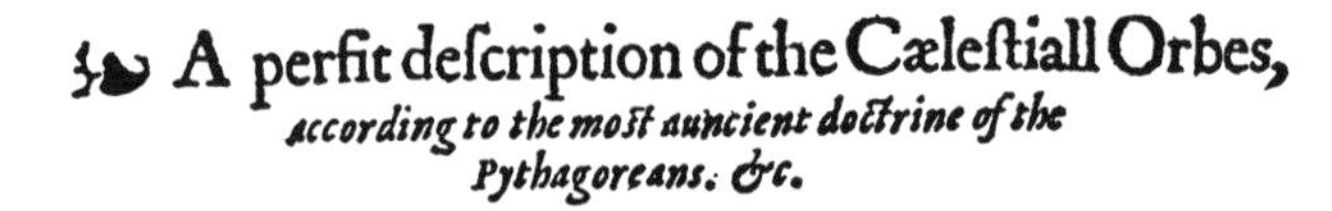

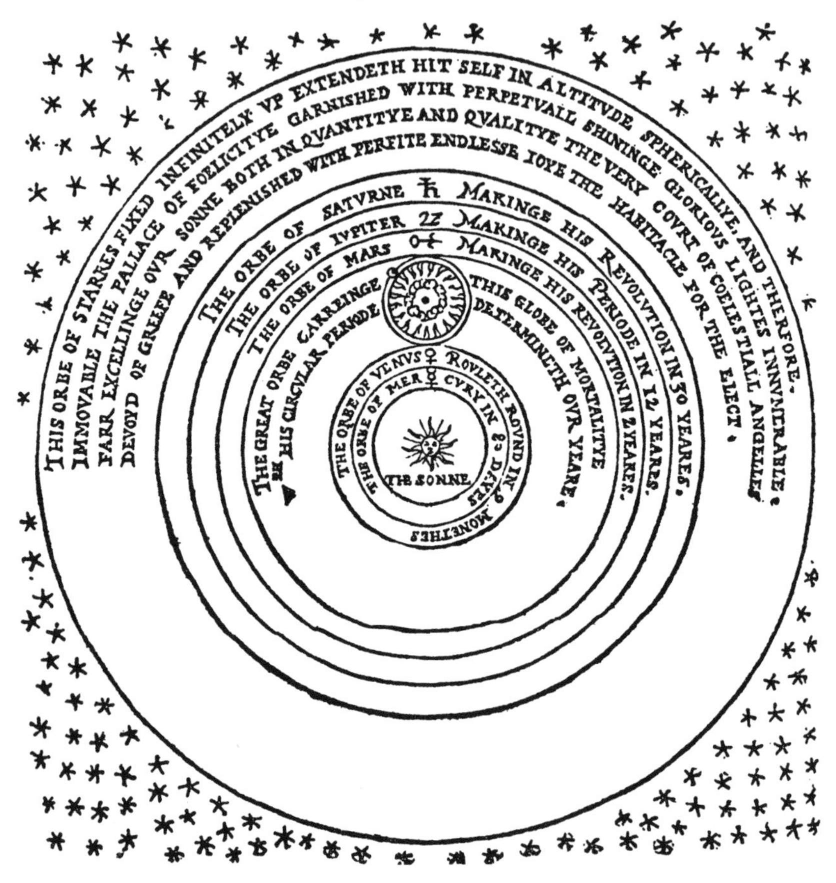

FIG. 1.2. Infinite universe of Thomas Digges (1576). The stars extend infinitely in space, limited only by the borders of the drawing. (Thomas S. Kuhn, *The Copernican Revolution*. Cambridge, Mass.: Harvard University Press, 1957.)

Currently, there is a dispute among cosmologists and astrophysicists about the nature and extent of dark matter, which is said to account for as much as 95 percent of matter in the universe. Some scientists question its existence, while others try to explain it in terms of known elementary particles. Perhaps as elusive as dark matter is dark energy, a repulsive, antigravitational force thought to account for the acceleration of our expanding universe.

Other Worlds, Other Life

The ancient Greek atomists were willing to entertain the idea of other inhabited worlds beyond ours, but the widespread acceptance of Aristotle's finite cosmos limited speculation about life beyond the Earth. The Moon was a possible location for extraterrestrial life, and several ancient writers imagined flying through the air and landing on the lunar surface. Even fictional travel beyond the Moon was unthinkable because Aristotle said that this region was filled with impenetrable, crystalline matter.

The Greek historian and biographer Plutarch (c. A.D. 46–120) was the most prominent early thinker to write about the Moon, its physical features, and possible inhabitants. Plutarch concluded that the Moon was similar to the Earth. Therefore, one way of learning about the Moon was to draw comparisons between it and the Earth.

If the Moon is similar to the Earth, we should expect to find some form of life there. Plutarch accepted the possibility of humanlike creatures living on the Moon. The lunar dwellers would be slight of build and obtain nourishment from plant life growing on the Moon. Plutarch went on to imagine a lunar creature looking at the Earth wondering if this motionless, cloud-shrouded body supported life.

Plutarch's use of the Moon as a platform from which to view the Earth and his willingness to draw analogies between the terrestrial and lunar landscapes and life forms were unusual for his time. Nevertheless, his way of thought eventually became a cornerstone of astronomy. Modern astronomical researchers assume that knowledge gathered on Earth is applicable elsewhere in the universe. For example, the same force of gravity that causes a stone to fall toward the Earth acts between the Moon and the Earth and also determines the orbital path of an extrasolar planet about its star.

Medieval philosophers who criticized Aristotle's finite universe also questioned his reluctance to accept the existence of life beyond the Earth. Theological issues, however, complicated the medieval response to Aristotle. A singular, finite cosmos provided a proper background for the history of Christianity. The Bible taught that the sacrifice of Jesus Christ for the sins of humanity (the Atonement) was a unique event that occurred on a unique body: the Earth.

However, Aristotle claimed that "there is not now a plurality of worlds, nor has there been, nor could there be."[1] First Aristotle confined God to a finite universe, then he ruled that God could not make other worlds in the future if He chose to do so. Aristotle placed limits on the creative powers of an infinite God.

In the fourteenth century, philosophers handled this thorny problem by announcing that an infinite God could, *if He wished*, create innumerable other worlds. They did not dispute God's ability to create a plurality of worlds, but few believed that He actually had done so. Therefore, the consensus was that humans lived in a unique world but that God possessed unlimited powers to create other worlds.

Hints of this medieval debate reverberate in twentieth-century space science. In 1975–1976 NASA sponsored a series of workshops on interstellar communication. The published proceedings of the workshops featured a foreword by the distinguished president of Notre Dame University, Father Theodore M. Hesburgh. Father Hesburgh, an enthusiastic supporter of the search for intelligent life, saw no conflict between his religious faith and the existence of other inhabited worlds. He explained his position in these words: "It is precisely because I believe theologically that there is a being called God, and that He is infinite in intelligence, freedom, and power, that I cannot take it upon myself *to limit what He might have done*."[2] The italics are Father Hesburgh's, but Christian philosophers first used this argument to challenge Aristotelian doctrines in the fourteenth century.

The possibility of other inhabited worlds demonstrated God's omnipotence, but it raised new questions for Christians. Had Christ died for humanity's sake or for the sins of all intelligent extraterrestrial beings? Was it necessary for Christ to travel around the universe sacrificing Himself repeatedly in each world? This matter was discussed in the Middle Ages, and it continued to trouble Christian theologians in the Renaissance. Some were reluctant to accept life on other worlds because it implied that Christ must redeem sinners living elsewhere in the universe. This was the view of the influential Protestant theologian Philipp Melanchthon (1497–1560).

Early modern scientists and thinkers who accepted the infinity of the universe also embraced the idea of inhabited worlds beyond the Earth. In the Renaissance, the idea of inhabited worlds claimed to have its roots in science, not theology, and the worlds were identified as planets circling other suns. This identification remains in force today.

The Italian philosopher Giordano Bruno (1548–1600) is one of the best known early advocates of the plurality of inhabited planets. Building upon the intellectual foundations laid by the ancient atomists and Copernicus, Bruno duplicated the Copernican heliocentric system and spread it throughout an infinite universe. Each of the countless stars in Bruno's cosmos was a sun with a group of inhabited planets orbiting it.

Bruno believed that life existed on the planets, but he did not speculate about the nature of extraterrestrial life. Instead, he rejoiced in all aspects of the infinite universe and celebrated the powers of a God who had created innumerable worlds. Unfortunately, Bruno also held philosophical and theological ideas judged heretical by the Roman Catholic Church. To mention one such idea, he questioned the divinity of Jesus Christ. In 1600 the Church ordered Bruno burnt at the stake for his unorthodox religious views.

The debate about the existence of an infinite universe filled with a plurality of inhabited planetary worlds continued after Bruno's death. Changes in the worldview introduced by the Copernican revolution persisted. In the seventeenth century, influential thinkers like Johannes Kepler, Galileo, and René Descartes joined the debate over life on other worlds.

The Principle of Mediocrity

An infinite universe filled with life-sustaining planets is one result of the revolution begun by Copernicus and developed by his contemporaries. There is another implication of Copernicanism that deserves attention in a study of extraterrestrial civilizations. This is the introduction of the so-called Copernican principle, or principle of mediocrity.

The principle of mediocrity is a twentieth-century concept that grew out of the Copernican revolution of the sixteenth century. Although Copernicus never mentioned the principle in his works, it was derived from his Sun-centered model of the universe. Copernicus eliminated the Earth's special status as the center of the universe. The Earth was simply one of several planets revolving about the Sun. If the Earth is a planet, then the other planets are similar to the Earth.

According to modern interpreters of the principle of mediocrity, our region of the universe is typical of the rest of the universe. There is nothing special about the Earth. The sequence of chemical reactions that nurtured life early in the history of the Earth, and the biological and cultural evolution of terrestrial life, happened elsewhere in the universe, leading to similar results. The evolution of extraterrestrial life follows a predictable path. It produces intelligent creatures who develop technologies of travel and communications similar to those found on Earth.

The principle of mediocrity covers all aspects of extraterrestrial life and civilization. Virtually every commentator on the subject of extraterrestrial life, and every scientific research program seeking intelligent alien life, operates upon this principle. Therefore, it is important to remember that the principle is an *assumption we make about the nature of unknown parts of the universe.*

The principle of mediocrity may be a reasonable and useful assumption, but it is an assumption nevertheless. In essence, we assume that the rest of the universe is much like the locale in which we live. Those who believe that extraterrestrial life exists also suppose that its history mirrors the development of life on Earth. Carl Sagan, the twentieth century's leading advocate of intelligent extraterrestrial life, admitted that the application of the principle of mediocrity to unknown regions of the universe is "essentially an act of faith."[3]

Superior Beings

Some modern scientists believe that life arose independently elsewhere in the universe. Living beings evolved progressively thereafter, and eventually intelligence emerged on other worlds. Over time this cosmic process raised the level of intellect to new heights enabling extraterrestrials to establish civilized societies. Although extraterrestrial civilizations resemble our own, they have reached stages of development that surpass the most advanced terrestrial civilizations. Carl Sagan put the matter bluntly when he wrote: "In fact, there is almost certainly no civilization in the galaxy dumber than us that we can talk to. We are the dumbest communicative civilization in the galaxy."[4]

The modern explanation of the superiority of extraterrestrial societies is that only older civilizations have developed to the point that they can communicate with us via radio signals. Once that is said, the focus is upon the superior characteristics of extraterrestrial civilizations and the promise they hold for the economic, social, and moral salvation of the human race. Younger alien societies, those at or below our level of civilized life, receive far less attention.

The idea of the superiority of celestial beings is neither new nor scientific. It is a widespread and old belief in religious thought. In philosophy, it began with Aristotle's view of the cosmos. Aristotle divided his universe into two distinct regions, the superior celestial realm and the inferior terrestrial realm.

The terrestrial region is a place ruled by change, growth, corruption, and death. In this inferior domain, men and women are born and die, and human affairs are in a constant state of flux and turmoil. By contrast, the celestial region, which reaches from the Moon to the outer limits of the universe, is eternal and changeless. If life of any sort exists in the celestial region, it will not be an extension of life found on Earth. Celestial life is more rarified and spiritual than terrestrial life.

Once again Aristotelian philosophy and Christian theology found common ground. Christians populated the celestial regions with God, the saints, angelic beings of varying ranks, and the souls of the dead (Fig. 1.3). These immortal celestial beings were superior to mortals, who inhabited the inferior terrestrial realm.

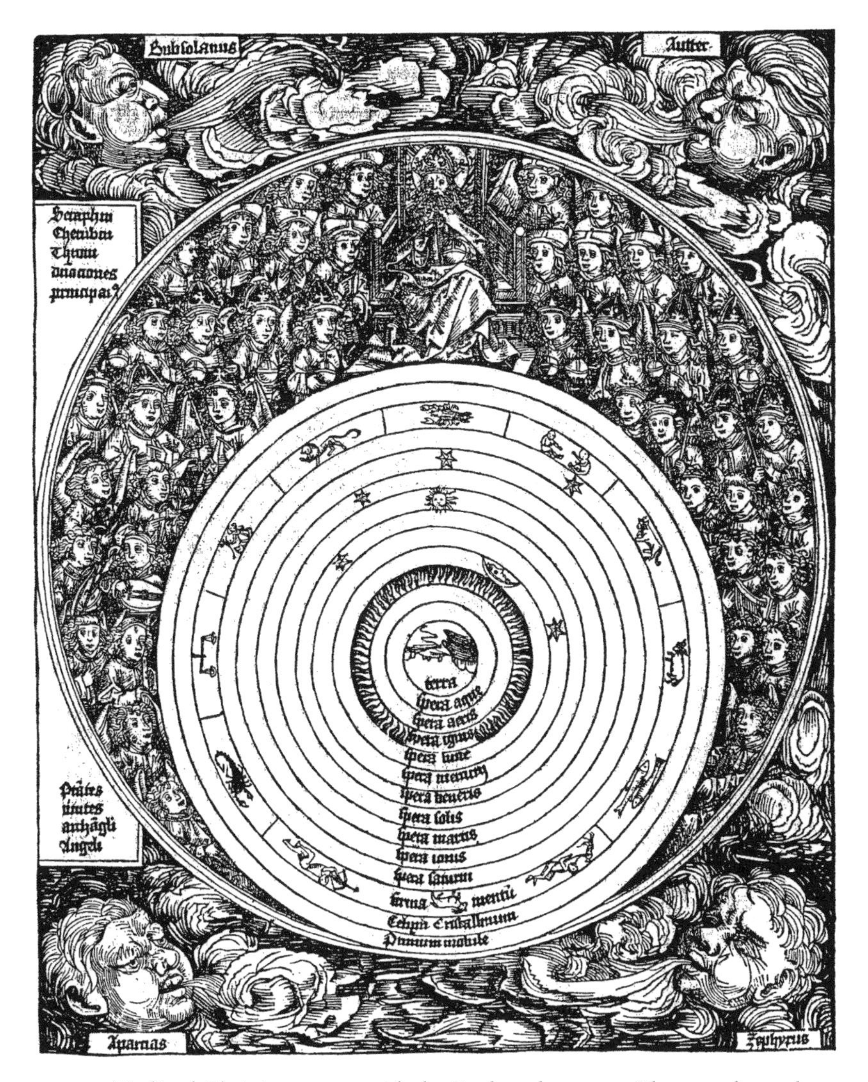

FIG. 1.3. Medieval Christian cosmos with the Earth at the center. The space beyond the stars is filled with God (seated at the top) and nine orders of angels. (Hartmann Schedel, *Register des Buchs der Croniken und Geschichten*. Nürmberg, 1493. Reprint New York: Brussel & Brussel, 1966.)

The Copernican revolution in astronomy erased the distinction between the celestial and terrestrial regions that was so carefully preserved by the pagan philosophers and the Christian thinkers who adapted his plan of the cosmos. According to Copernicus, the Sun, formerly a celestial body, was at the center of the universe. The displaced Earth orbited the Sun with planets once considered celestial. The motions Copernicus attributed to heavenly bodies called into question Aristotle's claim that the celestial region was filled with solid crystalline material. Soon thereafter, astronomers ruled out the existence of crystalline material, leaving empty space in its place.

By the end of the seventeenth century, the Newtonian picture of the universe prevailed. Most educated persons now accepted an infinite universe with heavenly bodies, governed by physical laws, moving through a void. Newton argued that gravitational attraction, acting on every piece of matter in the universe, caused the planets to maintain their orbital paths about the Sun. He accomplished this by providing detailed mathematical proof that accounted for the motions of the heavenly bodies. Newton published this great contribution to mathematical astronomy in 1687. Despite these fundamental changes in the conception of the universe, older notions lingered, especially the belief that creatures living on a distant planet were superior to the human species.

The Religious Impulse

So the idea of advanced extraterrestrial life emerged from a background of philosophical, religious, and scientific thought, with science the last ingredient added to the mix. Renaissance thinkers made extraterrestrial life a part of early modern science. This meant that old philosophical and religious notions of superior heavenly creatures took on new meanings in science.

Religious elements continue to adhere to the perception of extraterrestrial life even as we study it in the twenty-first century. Many modern scientists are not aware of the long and complex history, and the deep religious and emotional significance, of the idea of intelligent aliens. They are not dealing with scientific perceptions alone, but with old religious beliefs and philosophical concepts that underlie current scientific thinking.

Consider the case of Dr. Frank D. Drake (1930–), a pioneer radio astronomer and modern searcher for extraterrestrial intelligence. In 1960 he was the first to aim a radio telescope at nearby stars with the expectation that advanced civilizations in their vicinity might beam radio messages into space and thus to Earth. In 1981 an interviewer asked Drake what initially awakened his interest in extraterrestrial life. Drake's response was unequivocally clear: "A strong influence on me, and I think on a lot of SETI [Search for Extraterrestrial Intelligence] people, was the extensive exposure to fundamentalist religion." Drake, who attended the Baptist Church as a child, reported that many of his colleagues in the SETI movement "were either exposed or bombarded with fundamentalist religion."[5]

Drake reacted to his fundamentalist upbringing by turning from religious explanations of natural phenomena to scientific ones. He converted to the scientific way of thinking at an early age. Nevertheless, there are hints that his early religious training has had a lasting influence on him. In 1992, when NASA was set to launch its most ambitious project to search for extraterrestrial intelligence, Drake wrote a book on the scientific search for intelligence beyond the Earth. In this book, Drake announced that "immortality may be quite common among extraterrestrials."[6] Because Drake's notion of immortality had a scientific basis, he thought that extraterrestrial creatures might teach humans "how to live forever."[7]

Drake was not alone in believing that advanced alien life had achieved immortality. In 1981 NASA physicist Robert Jastrow claimed that older planets in the universe contained life "a billion years older and more advanced than man."[8] Scientists on these ancient worlds long ago discovered the secrets of the brain, united mind with machine, and created a race of immortal beings. Furthermore, these immortals had begun exploratory voyages throughout the universe. The immortal extraterrestrials discussed by Drake and Jastrow recall the centuries-old division between immortal celestial and mortal terrestrial beings.

Carl Sagan, like Drake, rejected traditional religion in his youth. In Sagan's case, it was Judaism, not Christianity. However, his biographer, Keay Davidson, argues that Sagan's insistence on the inevitability of extraterrestrial intelligence developed from his "quasi-religious belief in alien super-beings." Sagan was certain that these creatures were benevolent. They would help us solve current problems, like the spread of nuclear weapons and environmental pollution, by sharing their advanced knowledge with us. Davidson concludes that when viewed from a psychological perspective, Sagan's aliens "were secular versions of the gods and angels he had long since abandoned."[9]

In a 1970 lecture, Sagan recounted Soviet premier Nikita Khrushchev's boast that cosmonaut Yuri Gagarin saw no signs of angels or other supernatural beings in space when he orbited the Earth in *Vostok* in 1961. Sagan contrasted Khrushchev's critical remarks with the Apollo VIII astronauts who read from the Book of Genesis while in a lunar orbit. "[I]t is striking," Sagan remarked, "how space exploration leads directly to religious and philosophical questions."[10] A study of ancient and medieval thought shows that those questions are not new. The foundation of our thinking about extraterrestrial life draws upon older religious and philosophical ideas about life beyond the Earth.

The belief that superior beings inhabit the heavens appears in ancient myths, stories, and religious texts around the world. The modern aliens of science fiction and UFO literature, and the extraterrestrials of science, are secular versions of superior beings who originated earlier.

Why should so many people over the centuries believe in the existence of superior celestial beings? And, why did extraterrestrials become a legitimate part of modern science in the seventeenth century?

Psychologist Robert Plank answered the first question in his book *The Emotional Significance of Imaginary Beings* (1968). There he argued that humans have always had a strong emotional need to populate the heavens with imaginary beings. These beings often take the form of guardians who watch over humanity and serve as intermediaries between God and humans.

In support of Plank's thesis, compare the angelic visitations of an earlier time with UFO visits to the Earth today. The creatures in each case are superior to humans; their visits are sporadic, furtive, and disputed; and faithful members of a cult maintain their existence. If men and women in the Middle Ages believed that angels hovering nearby monitored their actions, a fair number of modern citizens believe that alien observers in hovering spacecraft are watching them. In the spring of 2000, *Life* magazine conducted a poll which indicated that 30 percent of the American population believed that aliens had landed on Earth.

Plank argues that each historical era spawns imaginary celestial beings appropriate to the times. Intelligent extraterrestrials are the imaginary creations of the scientific age. Even someone who rejects Plank's interpretation of the emotional origins of extraterrestrials might agree with his depiction of them as imaginary beings. Despite all their scientific trappings, the extraterrestrials discussed by scientists are as imaginary as the spirits and gods of religion or myth. We have no evidence for the existence of Martians, Jovians, or inhabitants of the recently discovered extrasolar planets. Using the principle of mediocrity as our guide, we believe that because intelligent life exists on Earth, it flourishes elsewhere in the universe. This is an acceptable belief but not proof of the existence of civilized extraterrestrial beings.

A distinguished historian of astronomy, Steven J. Dick, proposed an ingenious answer to the second question raised above: Why was the existence of extraterrestrials so widely discussed by early modern scientists in the seventeenth century? Dick maintains that by the seventeenth century, a mechanistic universe replaced the spiritual world of the Middle Ages. The new version of the universe was infinite and empty, except for the material celestial bodies—stars, planets, and comets—that traveled through it. Modern science demolished the medieval spiritual world and left the universe a vast lifeless vacuum.

Early modern scientists, Dick notes, created extraterrestrials to occupy the vacuum of space. They deliberately filled the void of the universe with beings capable of rational thought. Dick sees the projection of human intellect and reason into a region long occupied by supernatural beings as one of the great events in Western thought. He concludes that the idea of extraterrestrial intelligence modeled upon the human mind was as daring as any of the ideas introduced by Copernicus or Newton.

The hypotheses advanced by psychologist Robert Plank and historian Steven Dick to explain the acceptance of intelligent extraterrestrial life complement one another. Plank claims that humans throughout history have fulfilled an emotional need by imagining the existence of superior celestial beings. Dick argues that the destruction of the ancient Greek cosmos, along with its Christian spiritual world, deeply disturbed those who experienced it. Late Renaissance thinkers reacted by filling the desolate emptiness of the new universe with creatures possessing the equivalent of the human mind. These creatures lived social, cultural, and intellectual lives similar to those pursued by men and women on Earth.

CHAPTER TWO

Life on the Moon

> Once a comparison is [established] between the populations of the moon and of the earth, the judgment about similar things is the same. Since we see that the moon's spotted parts are civilized, we shall assign to the surrounding rough and mountainous regions wild and savage bands of thieves.
>
> —Johannes Kepler, *Somnium*, 1634

New Ways of Seeing, Old Ways of Thinking

The ancient astronomers and Copernicus made their great contributions to science before the invention of the telescope. They based their conceptions of the universe on astronomical data gathered by naked-eye observers. These observers used sighting and angle-measuring instruments that did not include magnifying lenses of any sort. The classic refracting telescope, consisting of a tube with an eyepiece at one end and a larger objective lens at the other, first appeared around 1609, more than sixty years after Copernicus's death.

The Copernican revolution did not originate in a new set of observations made with a novel scientific instrument. It was an intellectual revolution inspired by changes in the way early modern scientists thought about the structure of the universe. In any case, the telescope alone cannot supply crucial evidence for the

Copernican system. A viewer cannot see either a heliocentric or geocentric universe through the eyepiece of a telescope. Observations made with the help of a telescope are like any other sets of observations. Astronomers gather their data and then interpret it within the framework of existing scientific theory and practice. This complex process ends with a majority accepting a given view of the workings of the universe.

Historians divide observational astronomy into three periods. The first period, the era of naked-eye astronomy, dates from the earliest human observation of the skies and ends in 1609. This preoptical period included the work of Ptolemy (second century, A.D.) and Copernicus, two of the greatest figures in the history of astronomy. The second period began with Galileo's use of a telescope in 1609 to study the major heavenly bodies. Telescope makers devised new and more powerful instruments during the following three centuries, when optical telescopes ruled the astronomical sciences. Then, in 1931, Karl Jansky of Bell Telephone Laboratories detected radio signals coming from regions beyond the solar system. Jansky's discovery marks the beginning of the third period of observational astronomy, the era of the radio telescope. A radio telescope is essentially a large antenna, often shaped as a parabolic dish, used to detect, amplify, and analyze radio emissions from celestial sources. It is mounted so that it can be aimed at different portions of the sky.

Telescopes, both optical and radio, play an important part in the search for evidence of intelligent extraterrestrial life. However, astronomers must interpret the raw data collected with their instruments. The difficult process of interpretation yields ambiguous results that fuel scientific debates.

Galileo's telescopic observation of the Moon revealed large circular cavities on the lunar surface. Were these natural cavities of unknown origin, or did the inhabitants of the Moon build them? Late nineteenth-century telescopic observers of Mars saw a series of long dark lines on the surface of the planet. Were these lines generated by an observer's visual response to the natural Martian landscape, or were they evidence of a network of canals built by Martian engineers? Late twentieth-century radio telescope operators recorded periodic signals coming from outer space. Were these signals due to physical changes occurring in a distant celestial body, or were they coded messages from extraterrestrial beings in the universe?

In each of these cases, interpretation of the observational data depended upon the state of astronomical knowledge and current ideas about intelligent extraterrestrial life. Scientists can never escape the scientific, philosophical, and social assumptions that influence their best efforts to extract meaning from the observed world. In short, observations do not speak for themselves. Scientists shape their speech for them as they gain knowledge about the physical world.

Galileo's Telescope

Galileo Galilei (1564–1642), an Italian mathematician and philosopher, made the first systematic telescopic observation of the heavens in 1609–1610. Galileo used a twenty–power instrument he had constructed in his workshop. He first aimed his new telescope at the Moon and then at the stars and planets. The results of his pioneering work appeared in *Sidereus Nuncius (Sidereal Messenger)* in 1610. This treatise, the earliest publication in modern observational astronomy, inspired European astronomers to obtain telescopes and check the accuracy of Galileo's work. Meanwhile, Italian poets praised Galileo, comparing him to another famous Italian explorer of new worlds, Christopher Columbus.

Galileo announced the discovery of four moons orbiting Jupiter, the existence of stars invisible to the naked eye, the starry character of the Milky Way, and the nature of the lunar surface. He devoted a large part of his book to a careful survey of the Moon. Galileo illustrated his discussion of the lunar landscape with five drawings based on his telescopic observations. The Moon was an obvious choice for the first intensive telescopic investigation of a heavenly body. It is close to the Earth and it dominates the night sky.

In the old Greek cosmos, the Moon was a perfectly smooth spherical body traveling through the superior celestial region of the universe. Galileo's telescope disclosed a far different Moon. His had a rough craggy surface filled with chains of mountains, deep valleys, and plains. Galileo carefully recorded the movement of sunlight over the rugged lunar landscape. He calculated the height of lunar mountains and determined that some peaks were over four miles high, taller than any mountaintop in Europe.

Galileo used terrestrial analogies to describe what he saw on the Moon. He stated outright that a mountainous region of the Moon reminded him of the Bohemian landscape. Many scientists accepted Galileo's identification of the Moon with the Earth because it appeared in a book of observations drawn from nature and not in a polemical philosophical volume.

The boldness and novelty of Galileo's claims, and the sudden fame he gained, provoked criticism from some of his contemporaries. Astronomers found it difficult to duplicate Galileo's observations. Well–crafted telescopes of twenty or thirty power were not readily available. Furthermore, not every telescopic observer was as patient, persistent, and skilled as Galileo.

The use of the telescope in astronomy raised serious objections. Critics distrusted information gathered by this new instrument. The telescope appeared to distort physical reality. It made the Moon unnaturally large and transformed invisible stars into visible ones. In the past, scientific instruments measured length, weight, angular displacement, and the like. The telescope was a new kind of scientific instrument. It altered human perception of nature by revealing aspects of physical reality unavailable to ordinary sight.

With Galileo's help, the telescope soon gained acceptance among contemporary scientists. As the telescope's resolution and magnifying power increased, some astronomers thought that it might help them observe life on the Moon or planets.

Galileo's descriptions of an Earthlike Moon, and his discovery that Jupiter had a set of moons, helped to advance the principle of mediocrity. However, Galileo did not take the next step. He did not declare that life existed on the Moon. Galileo's reluctance to assume the existence of lunar life was due to his cautious approach to scientific speculation. He avoided controversial issues that might hamper the acceptance of the Copernican system. Throughout his career, Galileo was a staunch supporter of the work of the Polish astronomer.

Galileo made no mention of extraterrestrial life in *Sidereus Nuncius*. Twenty years later (1632), he wrote that he did not believe that herbs, plants, or animals similar to ours were propagated on the Moon, or that rains, winds, or thunderstorms occurred there. He said it was also unlikely that men inhabited the Moon. Galileo did not rule out the possibility of lunar creatures of a type unimaginable to us. He maintained that if advanced life forms existed on the Moon, they served and praised God.

Kepler and Galileo

Galileo sent a copy of *Sidereus Nuncius* to Johannes Kepler (1571–1630) soon after its publication. Kepler was an outstanding member of the new generation of Copernican astronomers. He had already transformed mathematical astronomy with his work on the shape of planetary orbits. For two thousand years, astronomers believed planets travelled along perfect circular orbits. Kepler used astronomical data gathered by the Danish astronomer Tycho Brahe to prove that the planetary paths were elliptical, not circular. This discovery was important to Sir Isaac Newton as he developed his comprehensive theory of the laws governing the motions of the heavenly bodies.

Galileo's lunar discoveries did not come as a complete surprise to Kepler. Ever since his early student days, Kepler had an interest in the Moon. In 1593 he wrote a treatise on the appearance of the universe as viewed by an observer located on the Moon. Kepler had read Plutarch's earlier treatise on the Moon, and he accepted Galileo's claim that the lunar landscape contained mountainous regions.

Kepler was well qualified to interpret lunar features revealed by Galileo's novel instrument. Earlier, Kepler had made naked eye observations of the lunar surface. In addition, because Kepler thoroughly understood the optical principles of the Galilean telescope, he suggested improvements to the instrument. When Kepler read *Sidereus Nuncius*, he learned that Galileo's telescopic observations confirmed many of his conclusions about the nature of the Moon.

There was one very important difference in Galileo's and Kepler's responses to the lunar data gathered in the new age of the telescope. Kepler, unlike Galileo, was willing to speculate about life on the Moon. Moreover, Kepler's speculations began more than ten years before Galileo announced his discoveries.

In 1610 Kepler published a little book in direct response to Galileo's telescopic observations. Its translated title reads: *Kepler's Conversation with Galileo's Sidereal Messenger*. Kepler noted that when he examined the Moon with a telescope, its surface showed features shaped by flowing rivers. Here was observational proof that there was water on the Moon. Kepler concluded that a mountainous Moon with deep valleys carved by rivers and a landscape dotted with lakes and seas must contain life. Since lunar mountains are higher than terrestrial ones, and lunar valleys deeper than those on Earth, Kepler assumed that lunar inhabitants are proportionally taller than humans. To continue the comparison, lunar technological projects are much larger than those undertaken by humans.

Lunar creatures must endure the intense rays of the Sun for long periods of time because the lunar day is equal to fifteen Earth days. Therefore, construction crews excavate huge circular pits on the Moon to provide shelter from the harsh environment. Underground lunar cities, not visible to our telescopes, lie beneath the surface of the large excavated pits. Lunar construction crews locate pasture lands and fields at the center of the pits to take advantage of the direct sunlight. When not tending their herds and crops, lunar herdsmen and farmers live comfortably near the shaded circular walls of the pits.

Massive lunar pits are not fanciful products of Kepler's imagination. Galileo published accurate sketches of prominent lunar features in *Sidereus Nuncius*. In a sketch depicting the Moon at last quarter, Galileo purposely exaggerated the size of a circular lunar cavity to emphasize illumination effects on the Moon's surface (Fig. 2.1). Kepler used Galileo's illustration as evidence that creatures capable of erecting large technological structures live on the Moon.

Kepler's interest in Galileo's discoveries went beyond the Italian astronomer's study of the Moon. When Kepler first heard that Galileo had observed four new bodies in the heavens, he hoped that they did not travel around a star. He was relieved to learn that the quartet of Galilean satellites circled Jupiter, a planet like the Earth. As Kepler explained to Galileo, four new planets circling a star would

FIG. 2.1. When Galileo drew this image of the Moon, he emphasized the size and circularity of a crater in the body's lower half. (Galileo, *Sidereus Nuncius*. Venice, 1610. Permission Wellesley College Library, Special Collections.)

support Bruno's doctrine of a plurality of worlds, an idea Kepler rejected. He believed in a finite universe dominated by our solar system.

Kepler carefully contemplated the meaning of the moons of Jupiter. God did not create the Jovian moons for human viewing because Galileo needed a telescope and a sharp eye to find them. In Kepler's mind, the intended uses of Jupiter's moons were self-evident. Our Moon shines for us. Similarly, the moons of Jupiter must shine for the benefit of the inhabitants of Jupiter. By a simple logical extension of his argument, Kepler determined that the other planets in the solar system had moons whose light illuminated their respective inhabitants.

Moonlight shines equally on all planets in the Keplerian solar system, but Kepler did not think that all planetary inhabitants were equal. At one point in his deliberations, Kepler raised a troubling question about the status of human beings. If some other planet is superior to the Earth, he said, then "we are not the noblest of rational creatures," nor are we "the masters of God's handiwork."[1]

In a not entirely convincing answer to this question, Kepler maintained that the Earth and its inhabitants were unique in the universe. God, he asserted, providentially situated the Earth at a proper distance between the vast wall of the stars and the central sun that warmed the universe. Kepler believed that God put humans in a preeminent position in the universe. They should be thankful to their Creator for placing them in this enviable location.

Kepler frequently mentioned God in his astronomical work. He did not make these references to appease religious authorities or satisfy the pious beliefs of his readers. When he entered the University of Tübingen, Kepler studied philosophy and theology in preparation for a divinity degree. By this time, the Catholic Counter-Reformation was underway and Protestant sects were quarreling among themselves over theological issues. Kepler, a Lutheran, was in the midst of these disputes.

Kepler never served as a Lutheran minister. After graduation from the university, he accepted a post as teacher of mathematics and astronomy at Graz, Austria. Thereafter he pursued a scientific career content that he could serve God, and come to understand Him better, by investigating His grand plan for the universe. In 1595 he wrote to a friend: "I had the intention of becoming a theologian . . . but now see how God is, by my endeavors, also glorified in astronomy."[2] Thus, God held a central place in Kepler's scientific thought. Extraterrestrial creatures may not have ranked as high as God, but they also had their place in Kepler's scheme.

Kepler's Dream

Kepler was interested in the Moon and its inhabitants throughout his life as a scientist. His most elaborate description of lunar life and culture appears in *Somnium (Dream)*, published posthumously in 1634. It contains his final and definitive statement on the motions, physical features, and flora and fauna of the Moon.

Kepler wrote *Somnium* to promote the acceptance of the Copernican system. He intended to advance it by comparing the relative motions of the Moon and Earth and demonstrating the probable existence of Earthlike creatures on the Moon. In his book, Kepler viewed the universe from the perspective of an observer on the Moon and drew parallels between the Moon and Earth. Inhabitants of the Moon, he wrote, would adopt a Moon-centered universe for the same wrong reasons that humans had once adopted an Earth-centered one. He declared that neither the Moon nor the Earth is at the center of the universe and that both bodies are in motion.

Kepler based his speculative account of advanced lunar life upon current scientific evidence. According to Kepler, three groups of intelligent beings live on the Moon. Subvolvans inhabit the near side of the Moon. On the far side, or hidden face of the Moon, live the Privolvans. Between them, in the high mountainous regions, is an unnamed savage race of thieves who raid the civilized settlements of the Subvolvans.

Nomadic Privolvans live on the harsh far side of the Moon. It is a place of open country, forests, and deserts where the nights are frigid and the days unbearably hot. Caves and grottoes, dug into the porous lunar soil, provide some protection from the heat and cold.

The daily lives of the Privolvans reflect Kepler's conjectures about their physical surroundings. Because we always see the same side of the Moon, observation of the dark side of the Moon is impossible with or without a telescope. The features of the far side of the Moon were unknown until October 7, 1959, when a Soviet spacecraft sent images of them to Earth.

The fortunate Subvolvans live in a more temperate climate than the Privolvans. Earthlight during cold nights and a cloud cover and rain during hot days moderate temperature extremes. Gardens and towns thrive in this mild Subvolvan environment.

Kepler discarded his earlier notion that the stature of Moon dwellers was proportional to the height of lunar mountains. In *Somnium* the Subvolvans are human sized, but there are large numbers of them. Kepler noted that the tower of Babel, Egyptian pyramids, Inca highway system, and Great Wall of China were not built by giants but by large organized teams of ordinary humans. A similar situation prevailed on the Moon. Lunarians joined together to work on big technological projects.

Kepler's telescopic observation of the near side of the Moon forced him to modify some of his earlier conclusions about it. He argued that the telescope confirmed his claim that water, in the form of seas and swamps, covers the lower regions of the lunar landscape. The telescope also disclosed a series of circular hollows or depressions on the Moon's surface. Modern astronomers say that these features are craters caused by meteoric impacts and volcanic action. Kepler interpreted the hollows differently. He concluded that they were artificial constructions built by civilized creatures who inhabit the Moon.

The hollows, he wrote, are the products of a rational architectural mind. The perfect circularity and orderly arrangement of the hollows reminded him of the cities and castles engineers build on Earth. He concluded that the Creator purposefully left the surface of the Earth and Moon in disarray so that their inhabitants could use technology to impose order upon a chaotic landscape.

Kepler claimed that his interpretation of the lunar surface was based upon telescopic observations and the axioms of optics, physics, and metaphysics. He felt confident that his rigorous approach made it possible for him to present new technical information about the nature and construction of the circular depressions. He determined that the radius of a typical hollow measured twenty-three miles and hypothesized that Subvolvan surveyors used a twenty-three-mile-long rope attached to a central stake to ensure the circularity of a hollow under construction.

When the surveying work was completed, the Subvolvans set to work digging a wide ditch within the circle and heaping the excavated soil outside to form a protective wall. Water from the interior of the circle drained into the ditch, leaving higher ground for farming. During the heat of the day, the ditch lost its water and became a roadway and a shelter from the Sun.

Science Fact and Science Fiction

Kepler's *Somnium* inaugurated a new literary genre. This genre joined the findings of science with literary inventiveness to depict the life and society of intelligent extraterrestrial beings. Literary critics claim that *Somnium* is the earliest known example of science fiction. They show a line of influence extending from Kepler to the pioneering science fiction writer H. G. Wells. Wells read *Somnium* before writing *The First Men in the Moon* (1901). Following Kepler's lead, he put his lunar civilization beneath the Moon's surface. In Wells's novel, lunar craters are part of a vast system of underground shafts extending for nearly a hundred miles toward the center of the Moon.

Apart from his role in science fiction, Kepler stands at the head of another literary genre. In this genre, working scientists depict details of the lives of intelligent extraterrestrial creatures. Here the lineage extends from Kepler to Giovanni Schiaparelli, Percival Lowell, Carl Sagan, Frank Drake, and to other modern scientists who speculate about the nature of civilized life on other worlds. Modern scientific narratives draw upon the available evidence about possible life in the universe and rely heavily upon the principle of mediocrity.

The principle of mediocrity claims that the Earth is essentially similar to the Moon and planets. Kepler emphasized the similarity and interchangeability of Moon and Earth in the Sun-centered system. In most cases, Kepler limited the implicit use of the principle of mediocrity to issues in physical and mathematical astronomy. He moved the principle of mediocrity in new directions in *Somnium* by enlarging the rule to include first the biological and then the social and cultural aspects of life. There were earlier precedents for modeling extraterrestrial life after terrestrial life, but Kepler carried these analogies into the realms of society and culture.

Kepler's account of life on the Moon deserves close examination. He is the first of a distinguished line of astronomers, physicists, and biologists to populate the universe with beings similar to humans and societies remarkably like their own. Speaking with the authority of science, these thinkers unwittingly inject their knowledge and experience of social and cultural matters into their descriptions of extraterrestrial life. Kepler's portrayal of lunar society owes as much to life in early seventeenth-century Europe as it does to the observations he and Galileo made with their telescopes.

Kepler spent most of his life in Austrian towns and small cities where he was close to the countryside. Even during his twelve-year residency in Prague, he was never far from rural scenes. Kepler transformed the familiar geography of Austria into his description of the physical features of the Moon. His depiction of the nearer side of the Moon (Subvolva) as a place of "cantons, towns, and gardens"[3] derives from the region in which he lived. The daily lives of the Subvolvans reflect the rural world he knew so well. Kepler's Subvolvans are farmers and herders who plant their crops and tend their flocks in the drained lands near the center of large lunar hollows.

Kepler's circular hollows, surrounded by earthen walls or ramparts, are suspiciously similar to the walled fortified cities of seventeenth-century Europe. In 1623 Kepler wrote to a friend that, with the help of a telescope, he had discovered "towns with round walls"[4] on the Moon. Three years later, Kepler was under siege in a walled city for two weeks when an angry peasant army blockaded Linz, Austria. He worked amidst the sounds and dangers of battle with soldiers quartered in his dwelling.

The construction of fortified cities demands a high degree of architectural and engineering skills. In the fifteenth and sixteenth centuries, the practitioners of military engineering gained special recognition as men who understood the mathematical theory, and the technical practice, of building fortifications able to resist enemy cannon fire. The ranks of those who excelled in this new art of warfare include such illustrious figures as Albrecht Dürer, Leonardo da Vinci, and Filippo Brunelleschi. Therefore, when Kepler had his Subvolvans plan and build circular fortified towns, he implied that the Moon dwellers were the equals of humans. They had the intelligence and technical prowess to undertake construction works that rivaled the finest engineering projects in Europe.

There were no terrestrial architectural models for lunar circular cities. Circular walls did not surround the fortified cities of Europe. Kepler was forced to improvise. He drew on contemporary surveying practices. Unlike later surveyors, men surveying land in Kepler's time did not use chains. They relied upon cords or lines for measurement. Therefore, his Subvolvan surveyors stretched long ropes as they measured off circular hollows on the lunar landscape.

The main point here is that when Kepler speculated about the technology of the lunar inhabitants, he transferred examples of current European technology to the Moon. This projection of terrestrial technology beyond the confines of the Earth is a prime example of the broadening of the principle of mediocrity. The underlying assumption is that the technology we see about us is the technology we are likely to find elsewhere in the universe. This assumption has profound implications for modern scientists who theorize about extraterrestrial technology as they formulate plans to search for, and communicate with, extraterrestrial intelligences. Kepler's brand of technological parochialism persists in modern depictions of alien technologies.

Mapping the Moon

Kepler described and interpreted the lunar landscape, but he never drew a map depicting the Moon's physical features. The mapping of the Moon dates to the emergence of modern cartography in the early Renaissance. Renaissance cartographers, driven by the needs of European explorers and maritime traders, produced accurate maps and charts in great quantities. The impulse that drove Renaissance cartographers to map land and sea also inspired astronomers to create detailed maps of lunar and planetary landscapes. The close relationship that existed between lunar and terrestrial mapping is suggested by a humorous remark Kepler once made to Galileo. He urged the Italian astronomer to draw a map of Jupiter, while he prepared one of the Moon, for sky travelers of the future.

Ancient geographers created maps of the Earth but not of the Moon. Near the close of the sixteenth century, William Gilbert (1540–1603), an English physician, sketched the first fairly detailed map of the Moon. Gilbert's naked-eye rendition of the Moon, unfortunately, was not published during his lifetime.

Gilbert drew a grid over the face of the Moon and introduced terrestrial terms to describe the physical features he observed. He noted continents, seas, bays, and islands, naming one of the lunar islands *Britannia*, in honor of his native country. Gilbert inscribed England's name on the lunar map at the same time that English cartographers were placing Virginia, England's first New World colony, on maps and charts of the Earth. Gilbert did not hesitate to describe features of the lunar landscape in terrestrial terms.

Galileo was the first to publish lunar maps based upon telescopic observations. His artistically sophisticated engravings of the Moon gave proof of what he had observed and indicated how others should interpret his observations. In keeping with his conclusion that the Earth and Moon were fundamentally alike, Galileo used terrestrial terms to describe the physical features he identified on the Moon. However, he never developed a formal lunar nomenclature. Instead, Galileo's lunar maps stand as a record of what he saw through the telescope when he observed the Moon.

The terrestrial model for the lunar landscape continued as astronomers used improved telescopes and new viewing techniques to help them produce better maps of the lunar surface. Johannes Hevelius (1611–1687) is prominent among the early describers of the physical features of the Moon. Using superior

FIG. 2.2. Map of the Moon published by Johannes Hevelius. (Johannes Hevelius, *Selenographia*. Danzig, 1647. Library of Congress, Prints and Photographs Division.)

telescopes, he observed the Moon over several years and sketched its features in detail. In 1647 Hevelius published the first authoritative set of lunar maps (Fig. 2.2).

Hevelius accepted the widespread notion that there was an ample supply of water on the Moon. He identified lunar "seas, bays, islands, continents, peninsulas, capes, lakes, swamps, rivers, plains, mountains, and valleys."[5] A moon that shared so many physical features with the Earth must also support life. Hevelius was convinced that Moon dwellers existed even though they could not be observed from Earth.

These early maps of the Moon are important for two reasons. First, they brought the Moon and Earth closer together in people's minds by emphasizing the many physical features they shared. This made it difficult to retain older Aristotelian conceptions of the Moon and encouraged speculation about lunar life. Second, lunar maps prepared the way for mapping the planets. When planetary landscapes were first studied, they, too, were compared to terrestrial landscapes and their physical features given terrestrial names. This in turn led to claims that higher life forms inhabited planets that, in other respects, resembled the Earth. The late nineteenth- and early twentieth-century maps of Martian canals drawn by astronomers Schiaparelli and Lowell were direct descendants of these first attempts to map the Moon.

CHAPTER THREE

From the Moon to the Planets

> A Man that is of Copernicus's Opinion, that this Earth of ours is a Planet . . . cannot but sometimes have a fancy, that it's not improbable that the rest of the Planets have . . . their Inhabitants too as well as this Earth of ours.
>
> —Christiaan Huygens, *Kosmotheoros*, 1696

Bishop Wilkins's Voyage to the Moon

The telescopic observations and writings of Galileo and Kepler inspired several seventeenth-century scientists to present astronomy to an educated popular audience. Bishop John Wilkins of England (1614–1672) was the first of these scientist popularizers. In his book *The Discovery of a World in the Moone* (1638), he defended the Copernican system, the idea of the plurality of worlds, and the probability of finding life on the Moon.

Probability theory originated in the seventeenth century when mathematicians in France, Italy, and Germany first analyzed games of chance, such as dice and cards. In England, Wilkins was one of a group of English thinkers who grappled with the new science of probability. He was among the first to realize the probabilistic nature of scientific knowledge and to distinguish between its degrees of certainty. Certainty could range from plain and clear physical evidence to belief and mere opinion.

Wilkins admitted that since we do not have firsthand knowledge of the Moon, we must remain ignorant of its inhabitants. We do not have certainty "or good probability" about their existence, he wrote. We can only "guesse at them . . . but we can know nothing."[1] Wilkins recognized that the probability of the truth of the Copernican system ranked above that of an inhabited Moon.

Wilkins was a founder of the Royal Society of London, England's oldest scientific institution. He was also a devout Christian and teleologist. A teleologist believes that God created the universe according to a design or plan that can be discovered in nature. Wilkins's acceptance of purpose in the universe led him to ask why a providential God made the lunar and terrestrial landscapes similar if He did not intend to put intelligent creatures on the Moon. His belief in teleology led Wilkins to consider the existence of lunar inhabitants.

Wilkins used the presence of lunar mountains to prove that humanlike creatures live on the Moon. He believed that a beneficent God put mountains on the Earth for a purpose. Terrestrial mountains, he declared, are not "so many heaps of rubbish left at the creation."[2] God artfully designed mountains for the beauty and convenience of the human race. Mountains tame the violence of rushing rivers, break the force of oceanic waves, strengthen the structure of the Earth, provide a refuge for wild animals, and offer humans a natural defense against their enemies. Therefore, when we find mountains on the Moon, we conclude that God put them there for similar reasons. Lunar mountains serve lunar inhabitants in much the same ways that terrestrial mountains make the Earth a safe and pleasant place for humans to live.

The intermingling of scientific and teleological arguments is understandable in a man who was an Anglican bishop and an astute scientific thinker. Wilkins's age accepted the compatibility of scientific and religious truths. The conflict between science and religion belongs to a later century.

Wilkins avoided giving a detailed description of the Moon dwellers and their way of life. Controversial questions about Adam and Eve and the salvation of humanity would arise if a clergyman in the Church of England discussed intelligent life on the Moon. Nevertheless, Wilkins raised the possibility of travel to the Moon and commerce with its inhabitants. He approached these novel topics in a cautious and rational manner, knowing that they might incite criticism, derision, or fanciful speculation among his readers.

Wilkins declared that in the earliest stages of world history, the Earth's great oceans acted as barriers between groups of humans. Only land travel was possible in this primitive age. The invention of oceangoing ships permitted people to journey freely over the face of the planet for the first time. In the new era of oceanic travel, worldwide trade and exploration flourished, bringing human communities together.

People living in the seventeenth century again found themselves isolated. This time they lacked vessels to carry them to the Moon. Wilkins admitted that there was no Francis Drake or Christopher Columbus ready to serve as guides to the New World in the heavens. He believed that humans, who had invented sailing ships in the past, would now find a way to travel to the Moon (Fig. 3.1). Once on the Moon, they could establish commercial relations with its inhabitants or possibly start a lunar colony. Colonization was on the minds of several European scientists who discussed the relationship between humans and extraterrestrials. Kepler predicted that the first nation to master the art of flying would be the first to colonize the Moon.

FIG. 3.1. Fanciful, bird-powered trip to the Moon. (Bishop Godwin, *The Man in the Moone*. London, 1638. R.A.S. Hennessey, *Worlds Without End*. Stroud, England: Tempus, 1999.)

Wilkins's proposal for a lunar voyage included a survey of the physical parameters of the trip. He determined the precise distance to the Moon, number of days of travel, nature of the attractive force that bound a space traveler to the Earth, limit of the Earth's atmosphere, diet of space voyagers, and the like. Wilkins assembled the first list of physical requirements for extraterrestrial travel. In later writings, he discussed a mechanical flying chariot, with spring-activated wings, to carry voyagers to the Moon.

It is fitting that a citizen of one of the great sea-going nations of Europe used a maritime analogy to depict space travel and looked upon Moon dwellers as future trading partners. Wilkins assumed that Lunarians would be similar to the peoples Europeans met in their travels around the globe. He did not believe that extraterrestrials represented a culture superior to European civilization. Hence, he never supposed that extraterrestrials might visit the Earth before humans visited them. Relying upon the New World analogy, Wilkins equated Moon dwellers with the primitive natives of America.

Descartes' Cosmic Model

Wilkins strongly defended the Copernican system. However, the Sun-centered model of the universe was nearly a century old when Wilkins published his book, and Copernicus had been largely concerned with solving problems in mathematical astronomy. Some saw a need for a new model of the universe, one that embraced recent advances in the physical sciences, astronomy, and philosophy.

In the mid-seventeenth century, the French thinker René Descartes (1596–1650) met the demand for a new cosmic model. Descartes was a philosopher and mathematician who made original contributions to physiology, the science of optics, and theoretical astronomy. He was the most famous scientist in Europe and was well prepared to offer the first comprehensive system of the universe since Aristotle outlined his idea of the cosmos. Sir Isaac Newton, who was Descartes' great rival, challenged Descartes' cosmos near the end of the century. However, Descartes' grand vision of the universe persisted well into the 1700s.

Descartes reserved the term "infinite" for God alone. Therefore, the universe was not infinite but indefinite in extent. Descartes rejected the void of the ancient atomists and filled every portion of his universe with moving particles (atoms). His mechanistic scheme reduced all celestial movements to the motion and collision of material particles.

In a universe packed tight with particles, the only possible motion is in large circular swirls, or vortices. Our Sun was the center of one of these vortices. The planets, caught up in the circling mass of cosmic particles, orbit the Sun.

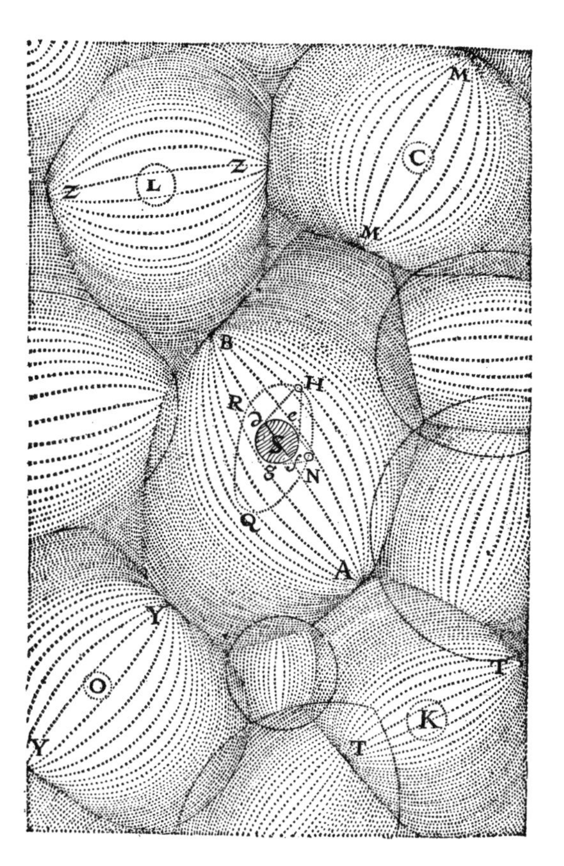

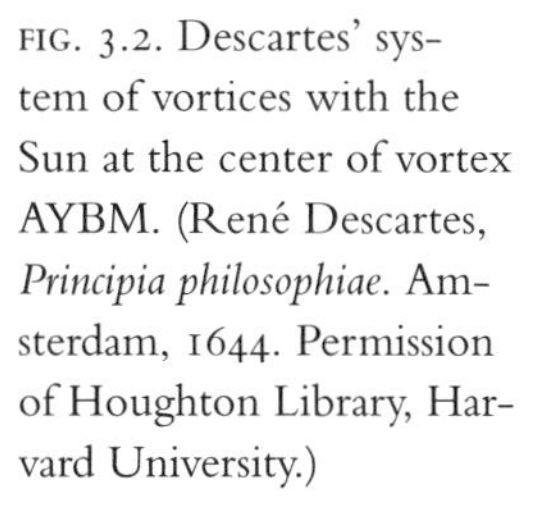
FIG. 3.2. Descartes' system of vortices with the Sun at the center of vortex AYBM. (René Descartes, *Principia philosophiae*. Amsterdam, 1644. Permission of Houghton Library, Harvard University.)

According to Descartes, the Sun is a star, and each star is at the center of a whirlpool of celestial matter. In the Cartesian system, it is reasonable to expect that the universe contains as many planetary systems as there are stars (Fig. 3.2).

Many planets revolving in distant vortices provide an excellent starting point for the doctrine of the plurality of worlds. Descartes, however, was slow to endorse pluralism. Early in his thinking, he hoped that improved telescopes might reveal animals roaming on the lunar landscape. By the end of his life, he limited himself to wondering if elsewhere in the universe there existed "innumerable other creatures of higher quality than ourselves."[3] Descartes deliberately avoided an extended discussion of extraterrestrial life. It was a subject certain to trigger a theological debate that would divert attention from the substance of his cosmological views. Descartes and Wilkins represented the cautious approach to the extraterrestrial debate prevalent in the first half of the seventeenth century.

By the end of the seventeenth century, the tide of informed opinion shifted toward the plurality of inhabited worlds. The new Age of the Enlightenment marked the beginning of modern times. Enlightenment thinkers hailed the victory of scientific rationalism over religious and supernatural thinking and the spread of the scientific outlook into fields as diverse as art, politics, and ethics. During this upheaval of traditional thought, the idea of a plurality of worlds was placed firmly in the scientific camp. Yet, despite the triumph of Enlightenment science, older, religious ways of thinking about extraterrestrials were not completely erased.

The Plurality of the Worlds: Fontenelle

No one better represents the new attitude toward extraterrestrial life than Bernard le Bovier de Fontenelle (1657–1757), the author of *Conversations on the Plurality of Worlds* (1686). Originally written in French, Fontenelle's book went through thirty-three editions during the author's long life. It was translated into English, German, Danish, Dutch, Greek, Italian, Polish, Russian, Spanish, and Swedish.

Fontenelle was not an original contributor to science. As secretary of the renowned French Academy of Sciences, he was in contact with the leading European scientists of the day. Fontenelle had a special talent for distilling the essence of current scientific thought into a philosophical work that was witty, engaging, and convincing. He openly drew upon Cartesian pluralism yet carefully evaded religious censors and critics. Fontenelle was not always successful in winning over the theologians. Their criticisms, however, were less important in an age that looked to science, and not to religion, for guidance.

The conversations in Fontenelle's book take place between a philosopher well versed in astronomy and physics and his charming, intelligent hostess, the Marquise. The two discuss the heavens as they stroll through the Marquise's country estate on five starry evenings (Fig. 3.3). Early in the book, the philosopher dismisses the teleological claim that God created all of nature for the exclusive use of mankind.

The participants in the *Conversations* do not explicitly rule out God's role in the cosmos. They simply replace Him with Nature and the laws of nature. On close inspection these laws are the same ones that rule Descartes' universe. Fontenelle's replacement of God with natural laws was not lost on his alert readers.

On the second evening of their conversations, the Marquise asks her scientist friend if there is life on the Moon. He uses a terrestrial analogy to answer his

FIG. 3.3. The philosopher instructs the Marquise about the nature of the universe. (Bernard le Bovier de Fontenelle, *Entretiens sur la pluralité des mondes*. The Hague, 1728. Universitätsbibliothek, Universität Basel.)

pupil. Telescopes reveal a Moon covered with seas, lakes, mountains, and valleys even though we are unable to observe lunar creatures. The similarity of the terrestrial and lunar landscape, however, is sufficient proof that life exists on the Moon.

The Marquise's searching questions about the nature of life on the Moon force her instructor to fall back on Kepler's underground lunar shelters. The philosopher explains that Moon dwellers can avoid the Sun's heat by retreating to cool shelters beneath the surface of the Moon. Underground cities and roadways lie hidden inside the Moon.

While discussing the inhabitants of the Moon, Fontenelle employs the maritime exploration analogy that Wilkins had used previously. The vague descriptions of Lunarians offered by her teacher do not satisfy the Marquise. She wants to see the Moon dwellers for herself, "to figure out what they're like."[4] The philosopher replies that she is as ignorant of the Moon's inhabitants as she is of Australian aborigines and yet she has never shown an interest in learning about the latter. The Marquise's rejoinder is that aborigines are humans who live on Earth. Lunarians are exotic creatures who inhabit another world. One might conceivably cross the Pacific Ocean but not the space separating the Moon and the Earth.

The philosopher responds that the inhabitants of the Moon and the Earth might establish communications soon. He reminds his companion that the natives of America never expected Christopher Columbus and his sailors to appear one day in great white ships spewing fire from their cannon. The Marquise immediately rejects this comparison. Of course ignorant American savages never imagined transatlantic travel. Enlightened Europeans, on the other hand, envision other worlds and contemplate traveling to them.

Thwarted again, the philosopher argues that recent aeronautic inventions could lead to lunar travel. A skeptical Marquise doubts the success of these early experiments in human flight. In desperation the philosopher replies that if humans cannot build workable flying machines, then the inhabitants of the Moon might develop the technological means to visit the Earth. "It doesn't matter," he says, "whether we go there or they come here."[5]

This statement marks a defining moment in the idea of extraterrestrial civilizations. Fontenelle introduces a fundamental shift with his suggestion that European technology may be inferior to lunar technology. The philosopher expands upon his new insight. He admits that "we'll be just like the [savage] Americans who couldn't imagine such a thing as sailing when people were sailing so well at the other end of the world."[6]

The remarkable suggestion that the Moon's inhabitants might be our technological superiors does not end this historic exchange. The Marquise asks shortly:

"Have the people on the Moon already come?"[7] This is another extraordinary response. It anticipates the so-called Fermi paradox. In 1950 Italian physicist Enrico Fermi asked a gathering of friends who were discussing extraterrestrial life: "Where are they?" Where are the representatives of the advanced civilizations that supposedly populate the universe? They should be in our midst by now.

Fontenelle's philosopher, aware of the differences between lunar and terrestrial atmospheres, notes that lunar travelers could not breathe our air. He imagines that explorers from the Moon might travel over the outer surface of the Earth's atmosphere. They might even angle for human specimens just as we fish for choice specimens in the sea. This possibility delights the Marquise. She hopes that lunar fishermen will catch her in their nets so that she can get a close look at her captors.

Fontenelle first reduces Europeans to the level of savages, and then to specimens in danger of being collected by curious visitors from another world. The tone throughout this discussion is playful, but Fontenelle pursues the argument in a deliberate and exhaustive manner. Finally, we are ready, along with the Marquise, to believe that the inhabitants of the Moon may possess technologies that surpass those of the most advanced nations of Europe.

Decades before Fontenelle wrote his book, authors of utopian and fantastic literature made satirical references to superior lunar societies. Fontenelle, however, is doing something different. He is writing a popular scientific work, not satirizing human folly. By carefully considering all the ramifications of the maritime exploration comparison, Fontenelle forces his readers to reassess their ideas about extraterrestrial civilizations.

At the close of their long conversation about lunar inhabitants, the philosopher is ready to introduce the Marquise to planetary life. Telescopic observations, which bolstered the case for an inhabited Moon, did not exist for the planets. Telescopes had improved since Galileo's time, but they could not resolve details of the physical features of Mars or Jupiter. Faced with the limitations of the telescope, the philosopher replaces it with another scientific instrument invented in the seventeenth century: the microscope.

The philosopher reminds his pupil that not all animals are visible to the naked eye. Microscopes have revealed an invisible world filled with strange forms of life. These microscopic creatures are as diverse and numerous as the visible ones that populate the Earth. Perhaps the planets contain as many microorganisms as the Earth.

The Marquise accepts this new argument, yet she admits to being "overwhelmed by the infinite multitude of inhabitants on all these planets."[8] Again Fontenelle breaks new ground. This time he joins the principle of mediocrity to his idea of the persistence and diversity of life. The result is a cosmos pulsating with life forms.

The comparison Fontenelle drew between Columbus and American savages is more convincing when applied to the Moon than to distant planets. Therefore, he discusses planetary life while developing another aspect of the maritime exploration theme.

In the seventeenth and eighteenth centuries, Europeans eagerly read travel accounts of strange flora and fauna, and exotic peoples, living in remote areas of the globe. Naturalists asked why life was so different in other parts of the Earth. The answer was that local environmental conditions caused the differences. The main determinants of culture were climate and geography. A cold climate engendered a hardy, vigorous race of people while a warm climate encouraged softness and laziness.

Fontenelle took the naturalists' explanation and made it into a general law: the greater the linear distance between two peoples, the greater the physical differences between them. Europeans and Africans differ to a lesser degree than humans and Lunarians. Humans, however, have more in common with Moon dwellers than they do with creatures living on distant Jupiter or Saturn.

The philosopher offers a further explanation of the diversity of life in the solar system. A planet's proximity to the Sun accounts for the distinctive features of its life forms. There are differences between Africans living in the hot tropics and Europeans living in moderate northern climates. Similarly, there are differences between the inhabitants of planets located near the Sun (Mercury, Venus) and those living on planets at a great distance from the Sun (Jupiter, Saturn). Whether we travel across the Earth, or across the solar system, climate determines the nature of life we meet locally.

Mercury presents a special problem. Because it is close to the Sun, it is hotter than the heart of Africa. Nevertheless, life exists on Mercury and on the slightly more hospitable Venus. Mars does not interest Fontenelle's philosopher. He turns his attention to Saturn with its rings and Jupiter with its moons and freezing temperatures. The philosopher imagines an astronomer on Jupiter viewing the Earth through his telescope and deciding that this tiny speck in the skies is lifeless.

The philosopher warns the Marquise not to expect to find inhabitants on the Sun or fixed stars. Each of the stars, she learns, is a sun at the center of a vast vortex in which planets invisible to us move in a regular fashion, driven by swirling celestial matter. The unknown planets circling in innumerable vortices, and the comets that move from one vortex to another, all support life.

Fontenelle brought life to the Moon, the solar system, and planets in far distant vortices by using physical evidence, carefully constructed logical arguments, and analogical reasoning. His creation of a biological universe marks a fundamental change in Western thinking about the cosmos and the plurality of worlds.

Huygens's "Probable Conjectures"

In 1656 the English scientific journalist and administrator Henry Oldenburg wrote to a friend, praising the latest developments in telescope design. If these improvements continue, he declared, we will soon be finding new countries in the heavens just as Columbus discovered new lands in America. Recent discoveries made by the young Dutch astronomer Christiaan Huygens (1629–1695) inspired Oldenburg's optimistic outlook. In 1655, using a telescope he designed, Huygens observed a satellite (Titan) circling Saturn. Four years later, Huygens published a book in which he claimed that a gigantic ring of solid material encircled Saturn. Modern astronomers have since determined that the rings are not solid. They are composed of particles of ice, varying in size.

Huygens was one of the most prominent physical scientists in Europe during the period bounded at one end by Galileo and at the other by Newton. An acute telescopic observer, he also ground lenses for telescopes and microscopes, invented one of the first pendulum clocks, and experimented with a crude form of internal-combustion engine. Huygens's research in mathematics led him to publish a treatise on mathematical probability. In the physical sciences, he made important contributions to our understanding of pendulum motion, the collision of material particles, and the wave theory of light.

Huygens devoted the last decade of his life to writing *Kosmotheoros* (1697), a book in which he presented a series of "probable conjectures" about life on other planets. This volume contained the first full statement on planetary life by an eminent scientist. It achieved immediate popularity and was read, translated, and cited as an authority throughout the eighteenth century. Thomas Jefferson had a copy of *Kosmotheoros* in his library, and John Wesley, the founder of Methodism, struggled with Huygens's pluralistic ideas as he developed his theological position on inhabited planets. Early in the eighteenth century, Tsar Peter the Great ordered a Russian language translation of *Kosmotheoros*.

The unstated principle of mediocrity exerted a strong influence on Huygens's speculations about planetary life. It led him to conclude that the planets, which share the Earth's motion about the Sun, must also share the Earth's life forms. He also assumed that similar cultural and technological conditions exist upon the Earth and the planets.

Huygens's study of the new theory of probability made him aware of the relativity of scientific truths. Early in the *Kosmotheoros*, Huygens admits that he does not have "certain knowledge" of life on the planets, nor can he assert that what he has to offer is "positively true." The best he can do is "to advance a probable guess, the truth of which every one is at his own liberty to examine."[9]

Huygens's sophisticated understanding of the conjectures he proposes lifts his book from the level of fantasy or fiction to the status of a scientific treatise. Huygens weighed his conjectures about planetary life as carefully as he did his hypotheses about the wavelike character of light. There is no set of attributes that define a work of science for all times and places. Each work must be evaluated within its historical context. Using that criterion, *Kosmotheoros* qualifies as a work of science by one of the great scientific thinkers of his day.

Despite his knowledge of astronomy and probability theory, *Kosmotheoros* reveals Huygens's commitment to teleology. He reminds his readers that science has recently discovered a universe that stretches far beyond the Earth and Sun. God created this vast universe and endowed it with purpose. However, its sole purpose was not to give pleasure to humans. God created other planetary worlds for the use and pleasure of other rational inhabitants of the universe.

Kosmotheoros, published twelve years after Fontenelle's *Conversations*, differed from its predecessor. Huygens concentrated upon rational beings, the thinking humanoids who inhabit the planets. For that reason, he gives a more detailed account of the anatomical structure, science, technology, and social lives of planet dwellers than Fontenelle.

Huygens differed with Fontenelle on another crucial matter: life on the Moon. Huygens reported that the telescope revealed lunar mountains, valleys, and plains, but no signs of life. The Dutch astronomer claimed that the best telescopes failed to find water in any form on the Moon. Moreover, the waterless Moon has no surrounding atmosphere and hence no air to sustain life.

Huygens studied the Moon through a powerful telescope, but he saw no signs of Kepler's artificial structures. The great circular features on the Moon's surface, he said, are much too large to have been built by lunar engineers. They are simply the results of natural forces operating on the Moon. After reaching that conclusion, Huygens hesitated. Would God create a large body like the Moon and leave it empty of life? He then admitted that plants and animals might exist there after all, but quickly added, "these are mere guesses, or rather doubts."

The planets, Huygens asserts, are a different matter. They are bodies very much like the Earth. Is it probable, he asks, that bodies that share the Earth's physical attributes are vast deserts covered with lifeless rocks and stones? Surely the Divine Architect would not have left the planets in a condition inferior to the state of the Earth? Thus, it is safe to conjecture that mobile living beings (animals) travel over the planets, feeding upon immobile life forms (plants). The Sun, which sustains life on Earth, also warms and nourishes the flora and fauna of the planetary worlds.

At this point, Huygens imagines a critic rebuking him for populating the planets with Earthlike organisms. He rebuts this criticism with a reference to life in the New World. God populated those distant lands with creatures that were familiar to the first Europeans who saw them. American animals, Huygens notes, have feet and wings like their European cousins, and they have similar hearts, lungs, guts, and reproductive organs. Europe is to America as the Earth is to the planets. Planetary life follows the organismic patterns established for terrestrial creatures.

Huygens determined that the planets have light and heat from the Sun and a sufficient supply of moisture. What sorts of plants, Huygens asked, are likely to thrive in a warm, bright, moist environment? The answer, of course, is plants similar to those growing on the Earth, plants that reach the loftiness and nobility of trees and propagate sexually. Huygens realized that given the similarity of environmental conditions, the variety of extraterrestrial animals also matches that of terrestrial fauna.

Planets overflowing with diverse flora and fauna are not complete if there are no rational creatures on them to enjoy the variety and beauty of their life forms. Huygens warned that rational creatures need not be exact replicas of humans. It is sufficient if they are similar to humans and endowed with reason. Therefore, rational beings must inhabit the planets. If they did not, the planets would be inferior to the Earth.

Huygens's determination to populate the planets with rational beings meant that the principle of mediocrity was expanded to include rationality. Huygens believed that reason on the Earth, Mars, or Jupiter was essentially the same. Along with reason, Huygens endowed planetary inhabitants with the senses of hearing, touch, smell, and taste.

According to Huygens, there are certain uses of reason that set humans clearly above animals. These include the contemplation of the works of God, the study of nature, and the improvement of the natural sciences. If it is true that the planets are not inferior to the Earth, then the intelligent creatures who inhabit the planets do not simply "stare and wonder at the Works of Nature." They do not merely view the stars; they study them in an organized fashion. In short, Huygens announced, "They have Astronomy."[10] The inhabitants of Jupiter and Saturn have an added incentive to cultivate the astronomical sciences because multiple moons circle their planets.

The study of the motions of the heavenly bodies requires the use of astronomical instruments made of wood and metal. Hence, we can assume that astronomers living on other planets have the support of carpenters and metal workers who manipulate the traditional tools of their crafts: saw, axe, plane, hammer, and file. Astronomers also need a knowledge of geometry, to make accurate measurements, and of writing, to record observations and make calculations.

Huygens's decision to equip his extraterrestrial astronomers with scientific instruments led him to ask if they use telescopes. Here he was caught between his love of telescopes, his conjectures about the nature of planetary life and intelligence, and his credibility as a scientific author. After giving some thought to the matter, Huygens decided not to put telescopes on the planets lest his critics ridicule him.

Huygens realized that his detractors could use New World comparisons to dispute his assignment of astronomy to planet dwellers. When the New World was first visited by European explorers, they found that its inhabitants were ignorant of science. Even in Europe, where modern science originated, a very small number of individuals did scientific work. Huygens stated that God in His infinite wisdom foresaw the coming of modern science. The Creator did not endow humans with scientific knowledge at their birth. Instead, He gave them the potential to develop the sciences over time. Therefore, we cannot deprive the inhabitants of the planets of the profit and pleasure of scientific research.

Huygens next confronted the issue of the anatomy of the rational creatures who inhabited the planets. He deduced that they had hands in order to operate instruments, manipulate pens and tools, and build houses and cities. They did not crawl around on their hands and feet but stood upright, a position that allowed them to use their eyes efficiently when studying the heavenly bodies. He decided that their overall size was close to that of humans beings but warned they did not share all human anatomical structures. Size was important to Huygens because, as he said, if his planet dwellers were "little Fellows,"[11] like rats and mice, they could not operate astronomical equipment.

Astronomy is constantly at the center of Huygens's speculations about advanced planetary life. It even dictates the nature of the social institutions found on the planets. The science of astronomy depends upon writing and the technical arts. Therefore, the planetary dwellers lived in settled communities where they did their astronomical work. In these communities, they engaged in trade and bartering and traveled about in boats. Huygens reasoned that if similar activities were carried out by New World savages, they must be common in planetary settlements.

Huygens concluded the first part of *Kosmotheoros* by reminding his readers what he had accomplished. He proved that inhabitants of the planets were rational creatures who possessed hands and feet, were versed in astronomy, studied geometry, lived in houses and cities, engaged in commerce, and were familiar with water transport. These ingenious and advanced creatures could easily replicate European technologies or surpass them with new discoveries.

The extraterrestrial world postulated by Huygens looks very much like Europe in the seventeenth century. We almost expect to meet an astronomer on Jupiter who is writing a book about inhabited planets entitled *Kosmotheoros*. Huygens took the principle of mediocrity to new lengths. He used it to spread reason and science throughout the universe and to create extraterrestrial technological civilizations modeled on life in Europe.

Huygens believed in the superiority of the astronomical sciences, the principles of teleology, and the importance of commerce for the prosperity of the maritime Dutch republic in which he was born. The same intellectual, religious, economic, and social forces that shaped Europe at the end of the seventeenth century also dominated extraterrestrial civilization.

Huygens lived in a country with a strong visual culture. In the arts, this culture was evident in the works of Vermeer and Rembrandt, two of the greatest painters in Europe. In the sciences, the Dutch emphasis upon the visual surfaced in the invention and use of two crucial optical instruments, the telescope and microscope. Huygens's fellow countryman Antony van Leeuwenhoek (1632–1723) became famous as a pioneer in using the microscope.

Dutch visual culture also shaped Huygens's scientific interests. He devoted his life to observational astronomy, the perfection of the telescope, and the study of the physics of light. However, when he depicted extraterrestrial civilizations, he exchanged his telescope for a mirror. He viewed European culture in a mirror and projected what he saw there onto the planets of the solar system.

Huygens's determination to put an astronomer on every planet leaves him open to the charge that his speculations about alien life never went far beyond his professional concerns and knowledge. However, his hypothesis is not that far removed from those modern scientists who believe that extraterrestrial civilizations construct radio telescopes.

The Maritime Analogy

The use of the maritime exploration analogy by Wilkins, Fontenelle, and Huygens may seem quaint or naive to modern readers. It is true that the analogy led seventeenth- and eighteenth-century thinkers to underestimate the difficulties involved in traveling to the Moon and the planets. However, comparisons between sea and space travel persisted and became part of the modern rhetoric of the space program in the mid-twentieth century. Proponents of the American space program revived the Christopher Columbus story to justify government support of missions into space (Fig. 3.4).

An early modern space reference to the Columbus theme appeared in *The Mars Project* (1952), a classic treatise on space travel published by the German-American space engineer Wernher von Braun. While setting out the overall

FIG. 3.4. Landing of Christopher Columbus in America. In 1994, this image was reprinted in *Where Next, Columbus? The Future of Space Exploration.* (Library of Congress, Prints & Photographs Division.)

plans for spaceflight, von Braun attacked the popular notion that a hardy band of adventurers could explore space using a solitary rocket ship. When Columbus began his historic voyage, said von Braun, he knew more about the Atlantic than we do about the heavens, yet he relied upon three ships, not one. A successful space program requires government backing of the sort that financed Columbus's voyages and the commitment to build a number of expensive spacecraft. Von Braun's proposed mission to Mars called for ten spacecraft, seventy crew members, and three "landing boats."

Space enthusiasts and publicists have repeatedly used the Columbus story to illustrate the benefits of space exploration. Critics said President Eisenhower hesitated to "hock his jewels,"[12] a reference to Queen Isabella, to finance U.S. space missions. President Kennedy, who differed with Ike and proposed the Apollo Moon project, called space a new ocean and urged America to become a spacefaring nation. Once space flights began, American astronauts were called modern-day Columbuses and planets became the equivalent of the New World.

One of the more bizarre uses of the Columbus legend occurred in 1960 when some early space explorers were concerned about "back contamination," possibly bringing extraterrestrial diseases to Earth on samples gathered on the Moon or planets. At that time biochemist Norman Horowitz compared the dangers of back contamination to the syphilis Columbus carried from the New to the Old World. The scourge of syphilis, Horowitz argued, must be balanced against the many benefits Europe gained from the discovery of America. Similarly, the risk of back contamination could be outweighed by the extensive knowledge acquired about life in the Solar System.

On October 12, 1992, the 500th anniversary of Columbus's arrival in the New World, the reinterpretation of Columbus's voyages reached its high point. NASA chose that day to inaugurate its search for intelligent extraterrestrials. The space agency directed powerful radio telescopes at the heavens in hopes of detecting messages transmitted by advanced technological civilizations. When, within a year, NASA was forced by Congress to halt its search, a participant said: "It was as if the *Niña*, *Pinta*, and *Santa Maria* had all been called back and mothballed within moments of pulling away from the docks."[13]

Columbus discovered a New World in America, and modern Americans commemorated his discovery by searching for new worlds in outer space. Thus, the circle was completed and the centuries-old maritime analogy kept alive.

CHAPTER FOUR

✦

The Ascension of Mars

> Astronomers were rewarded as they turned their eyes on Mars in the year 1877, when the orbits of Earth and Mars brought them into the best viewing position of the century. Observations made during this year by astronomers such as Asaph Hall and Giovanni Schiaparelli served to revive the myth of Mars as surely as the Roman Empire breathed new life into the Greek god Ares.
>
> —Jay Barbree and Martin Caidin with Susan Wright, *Destination Mars*, 1997

The Development of Planetary Astronomy

Huygens's *Kosmotheoros* was influenced by the models of the universe proposed by Copernicus and Descartes. However, eleven years before the appearance of Huygens's book, Sir Isaac Newton published his *Principia* (1687). The Newtonian universe, as outlined in his *Principia*, eventually displaced older models and stimulated the modern study of physics and astronomy, especially the investigation of planetary motion.

The Newtonian universe is the picture of the cosmos many of us hold in our minds today. The universe, said Newton, is an infinite void filled with massive bodies moving according to a set of mathematical laws. Newton emptied the skies of Descartes' particles and vortices. In their place, he put material bodies moving in a regular fashion under the influence of universal gravitation. The simplicity and precision of Newton's mathematical laws led many of his contemporaries to conclude that he had discovered the ruling principles of the cosmos, God's operating plan for the universe.

Newton wrote about the motions of the celestial bodies, but he was silent about the possibility of other worlds outside the solar system. He did not include these topics in his *Principia.* Newton finally broke his silence on the subject after prodding by the Reverend Richard Bentley. Bentley feared that nonbelievers might interpret the Newtonian universe as a godless mechanical system. Newton revealed his ideas about God's role in the origin of the universe and suggested that the stars might serve as centers of attraction for other planetary systems.

Those who accepted the Newtonian model of the cosmos did not share Newton's reluctance to comment on the plurality of worlds. By 1750 a large number of English and continental scientists, philosophers, literary figures, and popular writers discussed pluralism within the context of the Newtonian system. Pluralism and Newtonianism flourished together as the Newtonian world view was widely accepted during the late eighteenth and nineteenth centuries.

Newton's successful use of universal gravitation to explain the complex motions and orbital paths of the planets encouraged the development of planetary astronomy. Along with Newton's explanation for the mechanics of planetary motion, the planets themselves caught the attention of astronomers. The red planet Mars soon emerged as a favorite for telescopic observation and scientific speculation.

When Mars is at its brightest, it outshines every other body in the heavens except the Sun, the Moon, and Venus. Even though Mars is more distant from the Earth than Venus, the physical features of Mars are visible and those of Venus are not because of its constant cloud cover.

Christiaan Huygens was the first to investigate Mars in a systematic fashion. For forty years (1655 to 1694), he observed the Martian surface using improved telescopes he designed. In 1659 Huygens drew a rough map of Mars that was the first true drawing of the planet.

Later telescopic study of Mars led Huygens to conclude that Mars had a period of rotation of approximately twenty-four hours. In 1672 he drew another map of Mars that clearly showed Syrtis Major, which we now know is an elevated volcanic plateau, and displayed the planet's brilliant white south polar cap. Despite his long interest in Mars, Huygens had little to say about Martian life in the *Kosmotheoros.*

A number of European astronomers continued Huygens's groundbreaking observations of Mars. The noted English astronomer Sir William Herschel (1738–1822) contributed to the next major advancement in our understanding of the planet. Herschel was a superb builder of large telescopes, a tireless observer of the heavens, and a pioneering investigator of the structure of the universe.

Herschel modeled his telescopes on the new type of instrument invented by Sir Isaac Newton. The Newtonian reflector telescope uses a concave parabolic mirror to gather the light emitted by celestial objects. The light is then reflected through a small lens to the viewer's eye. A Newtonian reflector differs from the more familiar Galilean refractor telescope in which a viewer looks directly through a tube containing two different size lenses. It is possible to build huge Newtonian reflector telescopes, much larger and more powerful than refractor telescopes. Most major twentieth-century telescopes are variations on the Newtonian reflector.

Herschel discovered the planet Uranus, invisible to the naked eye, while observing the heavens with his powerful reflector telescopes. Uranus was the first new planet identified in historic times. Herschel also used his telescopes to make systematic surveys of the stars. His stellar observations spurred the development of stellar astronomy as a separate discipline in astronomical science.

Herschel also showed a strong interest in the existence of extraterrestrial life and civilization. Herschel's published and unpublished writings suggest that his ideas about inhabited worlds influenced his scientific work. It motivated many of his notable astronomical research projects, including his drive to design and build bigger telescopes.

Herschel announced the discovery of Uranus in 1781. One year earlier, he published a paper on the mountains of the Moon in which he made precise measurements of the heights of the lunar peaks. Herschel remeasured the heights of lunar mountains because he believed this new data revealed the strong probability, if not absolute certainty, that there was life on the Moon.

Herschel's unpublished astronomical notes show that he used his superior telescopes to search the Moon's surface for signs of lunar life. In May 1776, Herschel claimed he saw large areas of plant life on the Moon. If true, these lunar trees would be four to six times taller than terrestrial ones. Herschel hesitated for a moment and then plunged ahead into deeper speculation. It is possible, he wrote, that forests, lawns, and pastures exist on the Moon. Their presence would explain the changing colors of the lunar surface.

Two years after observing vegetation on the Moon, Herschel searched for signs of lunar towns hidden among the lunar forests. The circular lunar craters that fascinated Kepler also attracted Herschel's attention. Herschel claimed that the many circular formations visible on the Moon were towns erected by the

Lunarians. He urged astronomers to make a complete census of existing circular structures on the Moon so that they could identify any new ones built in the future.

Herschel's close inspection of the Moon's surface led him to classify the lunar circles into three categories: metropolis, city, village. Shortly thereafter, he found a canal, turnpike roads, and more vegetation on the Moon. Under Herschel's careful scrutiny, the Moon soon assumed the appearance of the English countryside.

Since Herschel found ample evidence of life on the Moon, he had little trouble populating the planets and their satellites. Although he wrote freely about the inhabitants of Saturn, Jupiter, and Uranus, Herschel showed a special interest in Mars. The twenty-four-hour rotation period of Mars, and the inclination of its axis, convinced him that a close relationship existed between Mars and Earth.

Because the Martian seasons were comparable to the Earth's, although longer, Herschel argued that the white polar caps he saw on Mars were mountains of snow and ice. Martian polar ice reflected sunlight, making it visible, and it partially melted during the warmer seasons of the year.

Herschel was the first to draw the dark markings seen on the planet. He suggested that the dark spots were oceans and the reddish tracts land. Further observation convinced him that Mars had an atmosphere, with clouds and vapors floating through it. Given the Martian seasons, polar caps, seas, soil, and atmosphere, Herschel concluded that Mars probably had an Earthlike environment.

Herschel's belief in extraterrestrial life extended well beyond the planets. He accepted the existence of a plurality of planetary systems. Each star was a sun with its own system of orbiting bodies, including inhabited planets. The light and warmth of their central sun sustained life on Herschel's extrasolar planets.

Returning to our solar system, Herschel made the extraordinary claim that the Sun was heavily populated. Solar creatures inhabited the Sun's cold interior. He maintained that only the Sun's exterior was hot (Fig. 4.1). Herschel's readiness to find life everywhere in the universe had its limits. He did not expect to find life on comets.

Herschel was not long alone in his search for extraterrestrial life. There was a renewed interest in the inhabitants of other worlds in the nineteenth century. This interest cut across the division of science and religion, with more scientists than religious thinkers accepting a plurality of worlds. Those who defended this plurality from a religious standpoint stressed the omnipotence of God and His power to create and populate multiple worlds. Scientists cited telescopic observations, or the latest chemical analysis of light emitted by a celestial body, to bolster their claim for the existence of extraterrestrial life. Religious and scientific adherents borrowed arguments and evidence from one another.

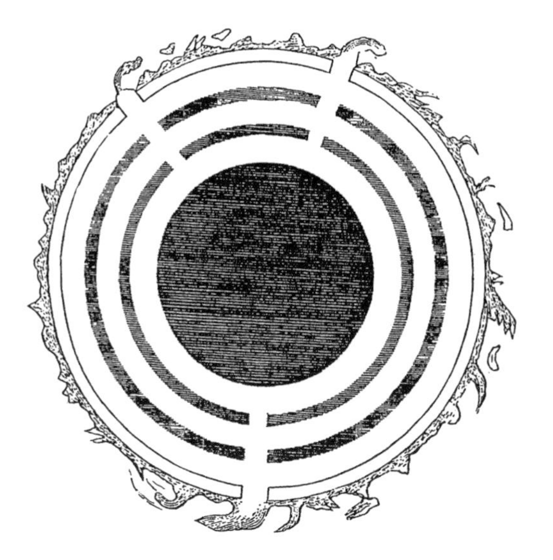

FIG. 4.1. Herschel's habitable Sun (1801). Solar dwellers are insulated from the Sun's heat by two protective layers. (Edward Dalziel, et al., *Half Hours in Air and Sky: Marvels of the Universe*. London, 1877. R.A.S. Hennessey, *World Without End*. Stroud, England: Tempus, 1999.)

The hopes of those who believed in life on other worlds in the nineteenth century centered on Mars. Professionals and amateurs alike turned their telescopes toward the red planet, observing its intriguing surface features. Before 1877 an army of observers sketched over one thousand drawings of the Martian landscape. No other planet attracted such close and prolonged attention.

Prime telescope viewing of Mars depends upon the relative motions of the Earth and Mars as they orbit the Sun. Mars can be as close as 35 million miles from the Earth as the two planets traverse their elliptical path. According to calculations made by astronomers, the years 1877, 1892, and 1909 would be optimum times to observe Mars.

The year 1877 proved to be a crucial year for the study of Mars. It marked the beginning of the modern phase of telescopic observation of the planet and initiated a period of intense speculation about Martian life. Astronomers aimed several new, powerful telescopes at Mars, including the giant twenty-six-inch-diameter refractor at the Naval Observatory in Washington, D.C. The Observatory's telescope, the largest of its kind, was under the direction of Asaph Hall (1829–1907). Hall intended to use it to search for the elusive satellites of Mars.

As early as the seventeenth century, astronomers discussed the possible existence of Martian moons, but no observer was able to find them. William Herschel searched for the moons of Mars unsuccessfully with his large telescopes. In August 1877, Hall saw the two Martian moons for the first time. He determined their orbits and had the honor of naming them (Deimos and Phobos).

A close reading of Hall's announcement of his discovery shows that he accepted the existence of intelligent life beyond the Earth. Two items in Hall's paper point to the topic of alien life. First, Hall includes a detailed description of how the two satellites would appear to a Martian astronomer. Second, he endorses the idea of a German astronomer to create a huge system of fire signals in Siberia in order to communicate with the inhabitants of our Moon.

Hall's mention of communication with lunar inhabitants recalls various proposals to communicate with Mars, Venus, or the Moon. The plans were to set large fires, or direct banks of electric lights with mirrors, to establish communication with intelligent alien beings. Several astronomers believed that Martians and Venusians were currently flashing bright lights aimed at the Earth.

In the same year that Hall discovered the Martian moons, other astronomers claimed they saw irrigation canals on the Martian landscape. Before assessing these startling developments, it is necessary to understand why observing a planet is more complicated than it may appear to be.

Through the Eyepiece

Soon after the invention of the telescope in the early seventeenth century, astronomers found that terrestrial atmospheric conditions drastically affected the telescopic study of the heavens. The gases that make up the Earth's atmosphere are constantly in motion, sending swirling currents of air in front of the telescope. These turbulent currents can blur critical details of a planet's surface.

Telescopic observers rarely have a sustained clear view of a planet as they peer through the eyepiece. Their view is intermittent with long periods of blurred confusion punctuated by brief moments of clarity. These conditions assume the planet under observation is in proper position for optimum viewing from Earth and there are no clouds in our skies.

Using more powerful telescopes with larger diameter lenses or mirrors does not solve the problems caused by atmospheric interference. Larger instruments simply enlarge the distortion caused by atmospheric conditions. At the end of the nineteenth century, astronomers debated the relative merits of large versus small telescopes for planetary study. Observers using small telescopes claimed they saw planetary features not evident to colleagues using larger instruments. This claim defied the law of optics.

Ultimately, it was the human eye that perceived the magnified image in the telescope, and the quality of vision differs from person to person. Apart from identifying common eye disorders, including color blindness, some nineteenth-century commentators drew an erroneous distinction between two kinds of observers. They claimed that persons with *sensitive* eyesight could discern the weak light from faint stars, while those gifted with *acute* eyesight were better able to detect the details of a planetary surface.

Unfortunately, the technology of photography could not resolve the observation problems troubling nineteenth-century astronomers. The photographic plates of the day were not sensitive enough to capture the image of a planet during the brief intervals when it appeared undistorted through the telescope. Instead, observers sketched what they saw in rare moments when the planet's image was distinct.

The talent and training of an astronomer affected the recording of a planet's features. Given two equally talented astronomers, one trained in engineering drawing and the other as a professional artist, a depiction of a planetary surface done by the engineer would tend to be precise, linear, and geometric. By contrast, the image recorded by the astronomer with artistic training would stress nuances of color, form, and shadow.

In the nineteenth century, the finished drawing required the intervention of yet another image interpreter, the engraver. This person engraved the inked plate used to print the image on the pages of a book or scientific journal. There were often substantial differences between the astronomer's firsthand sketch and the final printed image.

Other factors affect planetary observation. Sir William Herschel said it was an undisputed truth that "When once an object is seen with a superior telescope, an inferior one will suffice to see it afterwards."[1] Herschel's remark draws attention to a psychological dimension of telescope observation. Once an astronomer knows what to look for, and where to find it, perhaps a hitherto unseen object becomes easier to see.

When an eminent astronomer claimed to have observed a novel phenomenon or event, other observers would soon follow suit. In some cases, the scientific reputation of a leading astronomer would persuade others to endorse the initial discovery. In other cases, astronomers sharing similar theoretical outlooks would see the same things. These are not proof of fraudulent or deceptive practices by scientists. Instead, they demonstrate how psychological, social, and intellectual factors influenced planetary observation in the late nineteenth and early twentieth centuries.

All the factors affecting telescope observation became part of the Martian canal debate. This debate began at the end of the nineteenth century and lasted into the middle of the next century. Giovanni V. Schiaparelli and Percival Lowell claimed they observed a network of irrigation canals on Mars. Other astronomers failed to see the Martian canals. The parties involved in the canal debate disagreed about what their telescopes revealed and how to interpret what they saw and sketched.

Schiaparelli's Canals

The claim that Martians had built irrigation canals on their planet originated in the observations of the famous Italian astronomer Giovanni V. Schiaparelli (1835–1910). As a young man, Schiaparelli studied civil engineering at the University of Turin, where he specialized in architectural and hydraulic engineering. Schiaparelli learned how to design and construct canals, dams, sewers, aqueducts, pipelines, flood control systems, and other structures associated with the flow of fluids.

Despite his early professional training, Schiaparelli never practiced as an engineer. He briefly taught mathematics and then studied astronomy at observatories in Germany and Russia. In 1862 he became director of the Brera observatory in Milan, a post he held with great distinction for thirty-eight years.

Within a month after Hall discovered the two Martian moons, Schiaparelli decided to survey Mars and produce a new map of the planet. In 1877 Martian cartography needed a new naming system for the planet's prominent land formations. Although Schiaparelli promised that his nomenclature would not interfere with "the cold and rigorous observations of facts,"[2] the names he chose were not neutral. Schiaparelli designated dark bluish-green areas of Mars as bodies of water and lighter reddish-hued areas as land. In renaming the topographical features of Mars, Schiaparelli perpetuated the centuries-old practice of extending features of the terrestrial world into the extraterrestrial.

Of all the Martian landmarks Schiaparelli identified, none were more important than his *canali*. *Canali* is an Italian word that deserves special attention because it was translated into English as *canals*, that is, structures built by intelligent beings. The *Oxford English Dictionary* defines *canal* as "An artificial watercourse uniting rivers, lakes, or seas, for purposes of inland navigation, irrigation, or conveyance of water power." The Erie, Suez, and Panama Canals are all artificial waterways. In contrast, the English Channel is a natural waterway.

Unlike the English word *canal*, the Italian *canale* (singular) covers both artificial and natural watercourses. The English Channel becomes *il Canale della Manica* while the Suez Canal is *il Canale di Suez*. Schiaparelli called the thin dark lines he observed on Mars *canali di Marte* (canals of Mars). The ambiguity inherent in the Italian *canale* reflected Schiaparelli's own ambiguous response to what he saw in his telescope. Were *canali* artificial or natural? If they were artificial, this meant that Martian engineers were able to build and maintain an extensive system of waterways.

Schiaparelli observed *il canali di Marte* in 1877. Eight years earlier, *il Canale di Suez* had opened to oceanic ship traffic for the first time. The completion of the Suez Canal was an event celebrated throughout Europe and the Near East. Notable European and Near Eastern diplomats and political leaders were on hand for the occasion, and the Italian composer Giuseppe Verdi wrote the opera *Aida* to commemorate the canal's opening.

Schiaparelli, who had studied water-control engineering, was undoubtedly impressed by the construction of a waterway through the Egyptian desert. The Suez Canal ran for 103 miles, the longest ship canal in the world. It was called the greatest accomplishment of civil engineering in history.

The success of the Suez Canal stimulated large-ship canal construction around the globe. The great age of the ship canal coincided with a heightened interest in canals on Mars. As we will see, Martians did not build canals for inland navigation. The canals on Mars brought irrigation water to dry regions of the planet. Nevertheless, terrestrial and Martian canals represented the highest technological achievement of their respective cultures.

Schiaparelli was not the first to view or name the canal-like markings on the Martian landscape. Nevertheless, he went far beyond his predecessors in the extent and accuracy of his observations, the detailed precision of his maps of Mars, and the claim that the *canali* formed a complex network. He reinforced the idea of a wet Mars by emphasizing the presence of *canali* there. Water flowed in *canali* whether they were natural or artificial channels.

Schiaparelli drew comparisons between watercourses on Mars and the Earth when he interpreted observational data collected in 1877. Since 1830 observers of Mars had drawn a long line that ended in the so-named Solis Lacus (Lake of the Sun), Schiaparelli interpreted the line as a *canale* that drained its waters into the lake. By 1877 that line had disappeared from the Martian landscape. It had become invisible to observers using the best telescopes.

Schiaparelli had a ready explanation for the disappearance of the line. He wrote that changes in the "hydraulic regime of this region"[3] of Mars were similar to alterations the Yellow River in China had undergone recently. The Yellow River's water-control system included dikes, dams, and canals to restrain devastating flood waters and carry water to farm land.

• • •

The late nineteenth-century telescopic observations of Mars took place at the threshold of visual perception. Schiaparelli and the astronomers who supported or disputed his *canali* hypothesis speculated about the meaning of markings they glimpsed intermittently on the face of the planet. Viewings of Mars carried out within moments of one another yielded hand-drawn sketches that differed markedly.

Initially, Schiaparelli used a small 8.6-inch-diameter refractor to observe Mars, although he later moved to larger telescopes. Nineteenth-century astronomers disagreed about the best telescope size for viewing Mars. This debate coincided with questions about the quality of vision of the observer. It was common knowledge that although Schiaparelli was color blind, he studied a planet famous for its varied colored surface. His affliction, however, did not stop him from sketching an intricate web of lines he claimed to see on Mars.

Schiaparelli's interpretation of the markings on Mars as *canali* won him widespread support, as well as strong criticism. His supporters called him a modern Columbus who had discovered a new world on Mars. His critics challenged his interpretation of the evidence but never questioned his integrity as a scientist or his brilliance and dedication as an astronomical observer. Respect for Schiaparelli's scientific work often led astronomers to overcome their initial skepticism and accept the existence of the *canali.*

Schiaparelli's 1877 observations of Mars marked the beginning of the study of the planet by modern professional astronomers. His work brought Mars, and planetary astronomy, to the attention of the most skilled and experienced astronomers in Europe, Britain, and the United States. With Schiaparelli's map of Mars in their hands, or at least in their minds, large numbers of astronomers investigated the obscure markings on the planet.

Schiaparelli published a series of maps of Mars based upon observations he made between 1877 and 1890. He approached Martian cartography in the spirit of a civil engineer asked to lay out the plan of a building site on a plot of land or map a terrain. His maps were notable on two counts. First, Schiaparelli drew them with great precision. He approached the mapping of Martian *canali* as an exercise in the triangulation of the prominent surface features he observed. His survey of Mars included sixty-two fundamental points plotted with the aid of a micrometer. Schiaparelli's obsession with precise measurement gained him respect for thoroughness. It also raised the complaint that his "micrometric vision"[4] distorted his maps.

Schiaparelli's micrometric vision contributed to the second, and more controversial, aspect of his mapping of Mars. Critics often mentioned the diagrammatic and geometrical character of his maps. Schiaparelli boasted that his *canali* appeared to have been drawn using a ruler or compass. His critics charged that this precision was a result of the prior notions he brought to mapping Mars. He drew a scheme or diagram of the Martian surface that highlighted geometrical forms above all else. Schiaparelli's charts were more like engineering drawings than maps of a complex planetary surface viewed at a great distance under difficult conditions.

Schiaparelli never fully overcame his engineering education. He combined the technician's need for precision with the civil engineer's geometrical world view. His maps of Mars were bold and clear. They depicted *canali* that interconnected to form a planetwide hydrographic system. This system included *canali* that were 75 miles wide and 3,000 miles long (Fig. 4.2).

Schiaparelli and the English astronomer Nathaniel E. Greene both observed Mars in 1877. Shortly thereafter, each published detailed maps of the planet. Using a thirteen-inch-diameter reflector, Greene studied the planet from a favorable location on the island of Madeira. Greene's viewing conditions were excellent, and his telescope had a larger diameter than did Schiaparelli's. Greene produced a delicately shaded map filled with subtle distinctions of line and form. Greene's map showed no *canali* (Fig. 4.3). How could two expert observers study Mars at approximately the same time and produce different results?

Greene was not only an astronomer. He was a trained artist who had instructed the English royal family, including Queen Victoria, in the art of painting. The Reverend T. E. Webb, who evaluated the two maps in 1879, called

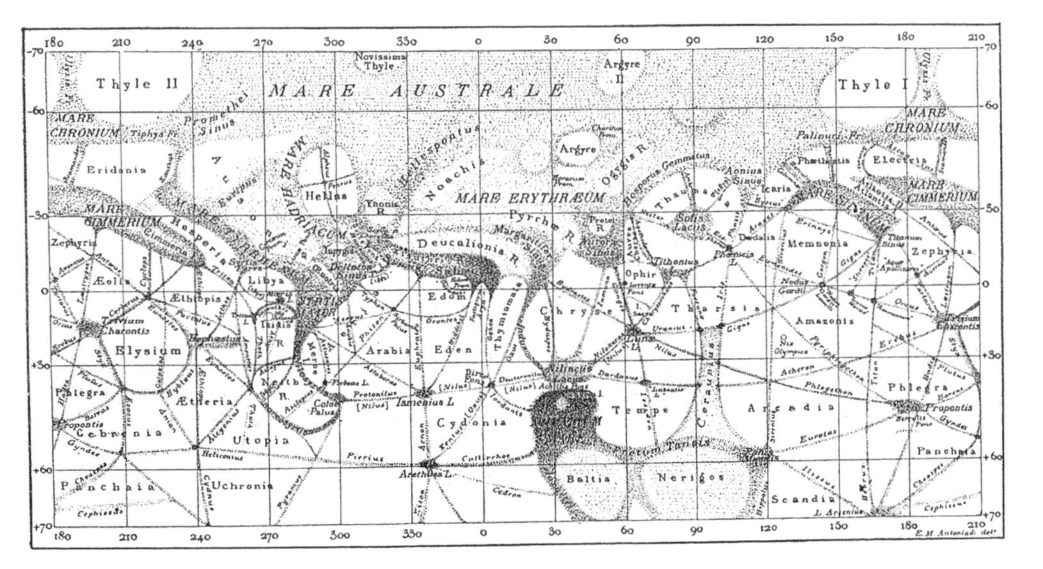

FIG. 4.2. Schiaparelli's map of Mars emphasizing the geometrical nature of the planet's canals. (Giovanni V. Schiaparelli, "La vie sur la planète Mars." *Bulletin de la Société astronomique de France*, 12, 1898.)

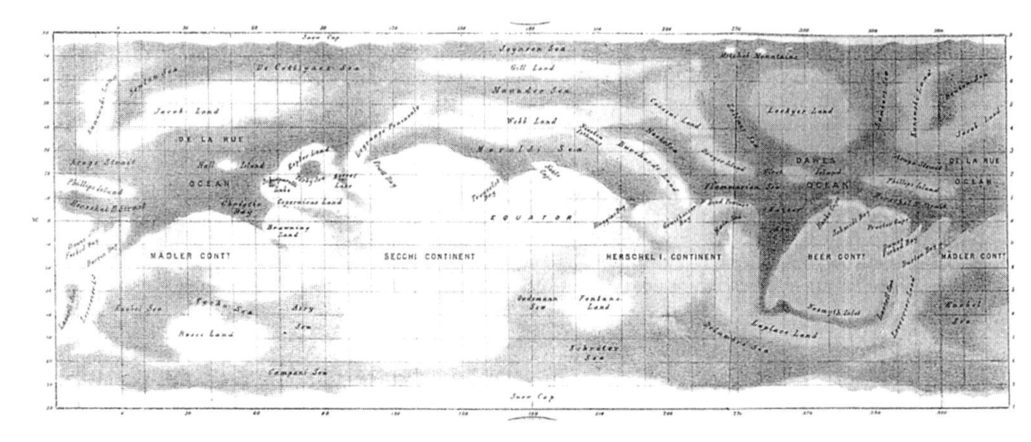

FIG. 4.3. Nathaniel Greene's 1877 map of Mars is dominated by large, indistinct masses. It features no canals. (Nathaniel E. Greene, "Observations of Mars, at Madeira in Aug. and Sept., 1877." *Memoirs of the Royal Astronomical Society*, 44, 1877–1879.)

Greene's effort a picture, or portrait, of Mars, and Schiaparelli's a carefully plotted and sharply outlined chart. In his review of Schiaparelli's map, Webb delicately reminded his readers that the Italian astronomer's vision was hampered by color blindness.

Greene attributed the hard and sharp lines of Schiaparelli's map to his drawing technique. The English astronomer suggested that either Schiaparelli's eye, or the eyepiece of his telescope, had a tendency to join a series of separate dots into the straight line of a *canale*. Schiaparelli responded that it was as impossible to doubt the existence of the *canali* as it was to question the reality of the Rhine River.

Shortly after publication of his map of Mars, Schiaparelli made a startling discovery. The *canali* were undergoing a process of "gemination" or doubling. Where at one time there was a single dark line, now there were two. The second line was parallel and equal in length to the original and set apart from it a distance of 210 to 420 miles. By 1882 twenty of the sixty *canali* Schiaparelli observed had doubled (Fig. 4.4). He concluded that gemination was not an optical illusion. He was absolutely certain he had observed a novel event on Mars.

Schiaparelli's latest observations brought new attention to the red planet. The level of controversy surrounding Mars rose as the number of astronomers searching for canals increased. Only a few observers were able to verify the process of gemination. Those who opposed Schiaparelli's *canali* hypothesis argued that the lines were not doubling. They said that the Italian astronomer suffered from eye fatigue. He was seeing double.

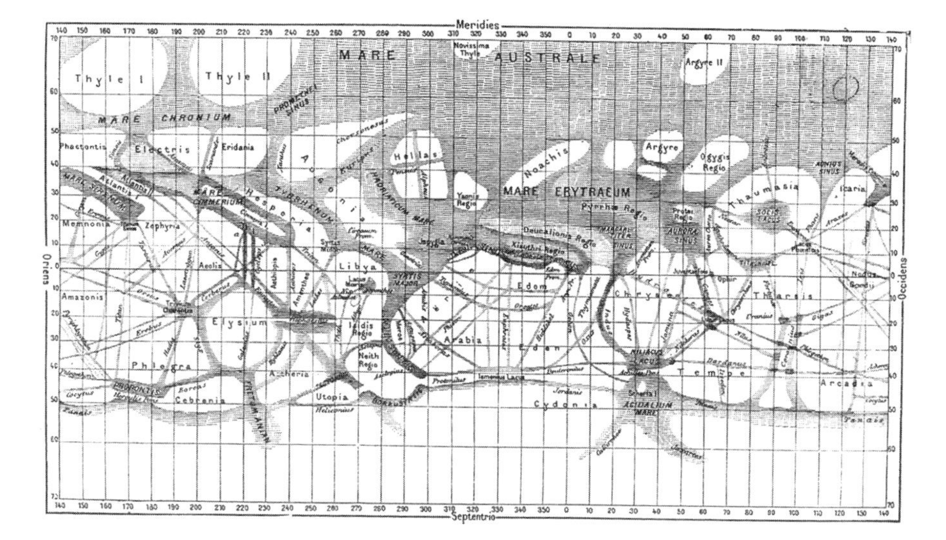

FIG. 4.4. Schiaparelli's map of Mars showing gemination, the doubling of canals. (Camille Flammarion, *La Planète Mars*. Paris, 1892. Permission Lowell Observatory Archives.)

The *canali* debate stimulated the imagination of astronomers. An English observer reported that gemination was an optical effect produced by mists hanging over Martian rivers at certain times of the year. A respected French astronomer announced that waters from an adjacent sea had recently inundated the huge Martian continent of Libya and that a canal ran directly across the northern Martian sea. The American astronomer William H. Pickering argued that the observed duplication of lines was due to variations in plant growth along the canals.

Any respect William Pickering gained from his vegetative theory of gemination he soon lost in Peru. In 1892 his brother Edward C. Pickering, director of the Harvard observatory, sent William to Peru to photograph stars and nebula. Instead of following his brother's instructions, William turned his telescope on Mars and telegraphed home that he had witnessed a Martian snowstorm as well as the melting of the accumulated snow. Sensational discoveries from Peru continued to mount, and Edward Pickering finally relieved his brother of his post.

As claims and counterclaims about the *canali* spread in the 1890s, the scientific dispute reached the general public. The dispute was picked up by science popularizers, writers of fiction, and sensationalist journalists. Interest in Martian *canali* reached such a peak in those years that a historian has likened the Martian canal furor to mass hysteria.

Martian *canali* emerged from scientific literature but soon entered a fantasy world of unbridled speculation. This included claims that intelligent Martians had built enormous structures on the planet, sent light signals, and made plans to invade the Earth. Inventors Thomas Edison, Nicola Tesla, and Guglielmo Marconi gave credibility to claims of Martian signals when they offered their technical advice to facilitate radio communication between the Earth and Mars.

• • •

The ambiguities of the Italian word *canali* cannot be blamed on Schiaparelli. The word originally referred to both natural and artificial watercourses. However, Schiaparelli exploited the double meaning of the word. If *canali* of any sort existed on Mars, that meant that there was water on the planet. Furthermore, Schiaparelli repeatedly drew *canali* as perfectly straight lines extending for hundreds of miles. He knew that only structures designed and executed by intelligent beings appear as if drawn onto the landscape with a straight-edge.

Schiaparelli saw himself as a disinterested observer, a dedicated collector of facts. His books, maps, and essays tell us otherwise. Schiaparelli's interpretation and visualization of the facts he gathered supported his deeper belief that creatures capable of completing great technological projects lived on Mars.

A year before his death in 1910, Schiaparelli confided to a friend that his eyesight had been deteriorating since 1890. Hence, he decided not to publish the results of any observations made after that year. However, Schiaparelli maintained an interest in Martian *canali* to the end of his life. His self-imposed ban on publication did not extend to general essays about the nature of *canali*. In 1893 and 1895, he published two articles in which he discussed the possibility and nature of intelligent life on Mars.

The Martian Engineers

Schiaparelli opens his 1893 paper with a description of the white polar caps on Mars. He compares them to the masses of snow and ice encountered by explorers in the Earth's polar regions. Martian ice and snow fit Schiaparelli's description of Mars as an aquatic planet. However, they conflicted with recent (1892) findings that the Martian polar caps were largely solid carbon dioxide, not all frozen water.

The seasonal melting of the Martian ice caps is central to Schiaparelli's interpretation of the landscape of Mars. Melting Martian snow and ice, he says, cause great inundations. Flooding is the main source of water for all the seas, oceans, lakes, swamps, and canals on the planet. Rain is very rare on Mars, and the planet's overall climate resembles that of a high terrestrial mountain.

When Schiaparelli turns to the *canali* in his paper, he continues to draw upon terrestrial analogies. Mars is essentially an aquatic planet like the Earth. Schiaparelli reports that every *canale* empties into a lake or sea, or intersects with other *canali*. He acknowledges that the geometrical nature of the *canali* recalls the work of intelligent beings, but finally settles on a natural, geological explanation for *canali*.

Schiaparelli next turns to the process of gemination. He admits that not all observers have been able to see the doubling of the *canali*. This is due to the need for proper observation of the phenomenon and to the seasonal nature of gemination. Schiaparelli notes that gemination coincides with the melting of the great northern Martian snow fields. The doubled lines appear, disappear, and reappear according to the seasons. They are not permanent geographical features of Mars, as are the *canali*.

The double lines Schiaparelli observed during gemination are so precisely aligned, he says, they resemble "the two rails of a railroad"[5] track. After making this tantalizing technological comparison, Schiaparelli immediately withdraws it. The distance between the "rails" is far too great for a railway system on Mars. Nevertheless, the precision of the lines suggests that they are the result of the work of intelligent beings. Schiaparelli's response to this suggestion is cleverly evasive: "I am very careful not to combat this supposition, which includes nothing impossible."[6] His use of the double negative permits him to avoid making a definite statement on the existence of life on Mars.

Schiaparelli argues that changes in vegetation over a large area, or the organic products of an enormous number of small animals, might be the source of gemination. He ends his 1893 paper by imagining an observer situated on the Moon studying the Earth. Changes on the Earth's surface caused by the blossoming of great fields of flowers, or by the agricultural operations of plowing and harvesting, would puzzle a lunar observer. Similarly, what we observe on Mars puzzles us. Are organic explanations the solution to the mysteries we find on Mars? Schiaparelli raises the question but leaves it unanswered.

Schiaparelli's 1893 effort is typical of his early treatment of the question of intelligent extraterrestrial life. His essay relies upon explicit terrestrial geological and geographical analogies, and it teases the reader with technological comparisons—railroads, artificial canals, agriculture, irrigation—that he withdraws.

In 1895 Schiaparelli returned to the question of Martian life and technology in a short paper that contained his most forthright treatment of the topics. He opens with a rudimentary lesson in Martian hydrography. The northern Martian polar cap rests upon a large continent, while the southern polar cap sits in the middle of a great ocean. The meltings of the frozen polar caps, therefore, have different consequences.

As the snow and ice in the southern polar region melt, water flows directly into the surrounding sea. By contrast, water from the melting northern ice cap floods the adjacent lowlands. Schiaparelli likens the seasonal flooding of northern Mars to the great tides that periodically roll over the lowlands of Holland or the northwestern coasts of Germany. He reminds his readers that the inhabitants of these European coastal areas react to the threat of tidal waters by erecting dikes.

The northern waters of Mars, Schiaparelli continues, are pure melted snow. They do not contain the mineral compounds dissolved in the saline ocean of southernmost Mars. Therefore, the fresh northern waters are crucial to life on the planet. Unlike the Earth, there are few clouds on Mars, and no rain, springs, or running water. The lives of Martian citizens, Schiaparelli concludes, depend upon their ability to capture and make the most effective use of the fresh northern water before it enters the southern sea. This is a job for Martian civil engineers.

Repeated northern floods have created large shallow valleys on Mars through which water flows swiftly during the warm seasons. Martian engineers, however, have intervened to control water flow with a system of strong dikes erected in the north. By this means, Martians regulate water flow southward. As the water moves through the broad valleys, it irrigates the cultivated land.

Imagine a wide shallow valley on Mars. On its broad sloping banks, engineers have built a series of canals that run parallel to the length of the valley. Some canals are situated higher than others because they are closer to the top of the sloping banks. Water in the canal network flows toward the south. In anticipation of this southern flow of water, Martians have built structures in the extreme southern regions of Mars to control the drainage of irrigation water into the sea.

When the spring floods begin, writes Schiaparelli, the Martian Minister of Agriculture orders the opening of the sluices in the north to fill the upper canals of the valleys with water. Water in the upper canals overflows the banks and then flows slowly down the sloping banks to the lower canals. The downward movement of water irrigates the broad cultivated zone located on the banks (Fig. 4.5).

Crops begin to grow in the irrigated zone when the dry Martian soil receives its first supply of nourishing moisture. Early in the growing season, there are two large cultivated zones, one along either side of the valley. At this time, an astronomer on Earth will observe a doubling of the dark lines on Mars. Gemination is nothing more than the appearance of two broad cultivated zones within the huge valleys. When the irrigating waters meet at the lowest point of the valley, bringing nourishment to crops growing near its center, a terrestrial observer sees only one line where previously there were two.

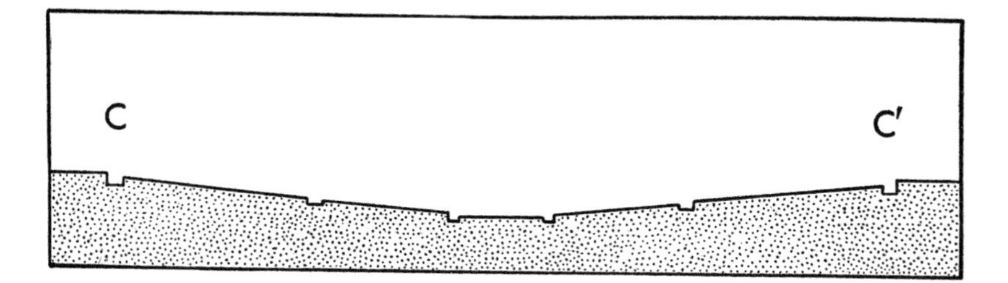

FIG. 4.5. Cross section view of Martian canals envisioned by Schiaparelli. The outermost canals (C, C′) are first to fill with water. At this point, a terrestrial observer sees a double line on Mars. As water overflows from C and C′ and runs down the slope to lower canals, the canals appear as a single line. (Willy Ley and Wernher von Braun. Illustrated by Chesley Bonestell, *The Exploration of Mars*. New York: Viking Press, 1956.)

Thus, a telescopic observer on Earth sees vegetation that flourishes on the sides of Martian valleys rather than the *canali* themselves. There are artificial canals on Mars even though our telescopes cannot detect them. What we see is the result of a massive technological and agricultural effort that turns areas thirty or more miles wide, and hundreds of miles long, into fruitful farmlands.

Schiaparelli the water-control engineer has been speaking to this point. Now Schiaparelli the social commentator takes over to speculate about Martian society. As the Italian astronomer shifts from the physical to the social sciences, he embraces the doctrine of environmental determinism. Schiaparelli believes that the physical environment in which people live determines their social institutions, political organizations, and values.

Martian society, according to Schiaparelli, is a "collective socialism" that developed from the need of its citizens to direct their energies against the common enemy, the harsh environment of their home planet. Schiaparelli "imagines a great federation of humanity," with each of the valleys constituting an independent state. Martians cultivate mathematics, meteorology, physics, hydrography, and structural technology to a high degree of perfection. International rivalry and wars are unheard of in this socialist paradise. Martian social life is centered on systems of dikes and canals that permit civilization to thrive in a hostile land.

In the last sentence of his 1895 paper, Schiaparelli writes: "I now leave to the reader the need to continue these considerations, and, as for myself, I descend from the hippogryph."[7] The hippogryph is a fabulous creature, part griffin, part horse, and thus Schiaparelli emphasizes the highly speculative nature of his remarks.

Schiaparelli sent a copy of his 1895 paper to an astronomer friend with these words written across the top: "Once a year it is permissible to act like a madman."[8] Rider of the hippogryph, or madman, Schiaparelli directed his speculations along the same scientific path he had been following since 1877. The maps filled with polar caps, seas, deserts, and precisely drawn *canali* were a prelude to the technocratic society he portrayed in 1895.

Schiaparelli's unrestrained thoughts on Martian engineers and socialism were not widely known in his lifetime. His paper was published in Italian, parts of it translated into French, but historian Michael Crowe first brought it to the notice of English speakers in 1986.

Critics of Schiaparelli pointed to a number of problems with his observations and interpretations of *canali*. Some noted that if an atmosphere existed on Mars, it did not contain water vapor. The glistening white polar caps were mainly crystals of frozen carbon dioxide, not snow. Others argued that *canali* and gemination were simply optical illusions. An observer of Mars, operating at the edge of visual perception, could easily transform a series of small dark spots on the planet's surface into continuous dark lines.

• • •

Whatever its limitations, and however short its life, Schiaparelli's theory offered scientists a way of thinking about Mars when the planet first came under sustained scrutiny through telescopes. The *canali* theory played a legitimate role at a crucial time in the development of planetary astronomy.

Schiaparelli's conception of Martian life and technology reveals a trend that began when early scientists first commented on extraterrestrial life. Just as Huygens populated his planets with rational beings dedicated to astronomy, so did Schiaparelli, a trained civil engineer, populate Mars with engineers working on gigantic water projects.

Yet another trend is clear in Schiaparelli's speculations. Martians are not only intelligent; they are superior to humans. Others hinted at the superiority of extraterrestrials, but Schiaparelli helped to popularize the idea. Schiaparelli's Martian engineers constructed a canal system that dwarfed the Suez Canal and made their planet habitable. No example of terrestrial technology was comparable to the canals on Mars.

A contemporary of Schiaparelli's, the French astronomer and popularizer Camille Flammarion, offered an explanation for the superiority of Martians. He maintained that because Mars was older than the Earth, its inhabitants had evolved to a higher state of intelligence. Percival Lowell, who carried on the canal debate after Schiaparelli's death, perpetuated this depiction of Mars as an aging planet inhabited by superior life forms. And the debate begun by Schiaparelli became even more spirited under Lowell's guidance.

CHAPTER FIVE

Percival Lowell
Champion of Canals

Prof. Percival Lowell is certain that the canals on Mars are artificial. And nobody can contradict him.

—Clipping from unidentified newspaper (summer 1905)

The Orientalist

On May 19, 1910, less than two months before his death, Schiaparelli publicly stated that natural forces could account for the dark lines seen on Mars. However, he went on to suggest that someone assemble all evidence related to the existence of intelligent life on the planet. In the concluding paragraphs of his final communication on Mars, Schiaparelli mentions his admiration for the work of Percival Lowell. This praise for Lowell raises doubts about Schiaparelli's acceptance of a natural explanation of the *canali*. Lowell was an outspoken defender of canals built by Martians throughout his scientific career (1894–1916).

Percival Lowell (1855–1916) was the most powerful champion within the scientific community for the idea of intelligent Martian life. His claim that the Martian landscape included a global irrigation system influenced the conception of Mars held by scientists, government officials, and the general public well into the second half of the twentieth century. Unlike Schiaparelli, Lowell wrote popular books and magazine articles and lectured widely on Martians as canal builders. Lowell was an energetic and effective publicist for his views on Martian life.

Some writers have called Lowell a newcomer to astronomy, an outsider, and even an amateur. When Lowell began his scientific career, entrance into the profession did not require an advanced degree in astronomy. A number of distinguished early twentieth-century astronomers, including directors of major observatories, never received advanced training in astronomy. Lowell's credentials as an astronomer were not unusual for his times.

Lowell studied mathematics at Harvard University under America's premier mathematician, Benjamin Peirce. Peirce fully expected his brilliant student to succeed him as a professor of mathematics. Lowell had different plans for the future. After spending a year abroad, and the next six years attending to the business holdings of his illustrious and wealthy family, Lowell left America to study the Far East. In 1882 Lowell had attended a lecture on Japanese culture by zoologist Edward S. Morse. Morse's lecture inspired Lowell to travel to Asiatic countries recently opened to the West.

Percival Lowell was one of a number of socially prominent Americans who visited East Asia in the late nineteenth century. Most of these observers left travel accounts of their adventures. Lowell did more. He was among the first Westerners to define the essence of Eastern civilization and show how it differed from Western civilization. Between 1883 and 1893, Lowell spent most of his time in Japan and wrote four books on Oriental life and culture. Contemporary Orientalists regarded Lowell's books as serious assessments of the Oriental mind and personality. However, today's Asian scholars find them written with a strong bias toward European culture.

There are definite connections between Lowell the Orientalist and Lowell the astronomer, and between Orientals and Martians. In the 1880s, the Eastern mode of thinking and way of life were ripe for analysis by Western observers. Lowell visited Japan and Korea and quickly plumbed the depths of the Oriental mind, at least to his satisfaction. Using a telescope, he next visited another set of exotic beings: Martians who built planet-wide irrigation canals. In both instances, Lowell's ideas were influenced by the writings of the English thinker Herbert Spencer (1820–1903), a social philosopher highly regarded in America. Spencer believed that neither state control nor social reformers should intervene to influence the evolutionary progress of society.

Following Spencer's lead, Lowell used scientific and material progress to measure the level of civilization attained by a people. He was convinced that he was scientifically investigating the Oriental mind. This led him to conclude that the West, which vigorously developed science and technology, was superior to the stagnant East, which failed to do so. Oriental civilization was trapped in the early immature stages of cultural and technological evolution described by Spencer. Employing a striking astronomical analogy, Lowell compared Japan to the barren and dead Moon. According to Lowell, a crucial distinction between East and West was the impersonality, or lack of the sense of self, of Orientals and the pronounced individualism of Westerners, notably Americans. Lowell attributed the impersonality of Orientals to their deficient imagination. Instead of identifying imagination with the arts, Lowell believed it was essential to science and mathematics. Imagination, he contended, was far more crucial for scientific investigation than observation.

The priority of insight over eyesight was central to Lowell's conception of science and to his interpretation of the dark lines seen on Mars. Lowell believed it was not enough to observe Mars and make accurate drawings of what one saw. One must boldly go beyond the observations and propose imaginatively conceived hypotheses and theories. Astronomers were not a band of technicians. They were generalists who aspired to become philosophers.

Another piece of evidence from Lowell's stay in Japan is relevant to understanding him as a speculative thinker and theorist. Basil Hall Chamberlain, a fellow Orientalist, remarked that Lowell was so certain of the truth of his theory of Oriental impersonality that he refused to entertain any evidence to the contrary. Chamberlain wrote to a mutual friend that Lowell argues deductively "from some general notion . . . and then bend[s] the facts to suit the preconceived idea, seasoning the whole with verbal fireworks."[1] In later years, Lowell's scientific opponents echoed Chamberlain's criticisms.

If the imagination is supreme, as Lowell believed, then how can we differentiate between a reasoned scientific theory and uncontrolled speculation? Lowell was aware of the dangers of a giddy imagination. Nevertheless, he found it difficult to accept new data, reevaluate old observations, defer to the expertise of others, or modify his theories.

Lowell established a mode of theorizing in the Orient that he carried with him into his scientific work. He also showed an interest in astronomy during his lengthy visits to the Far East. This interest was strengthened by Spencer's claim that astronomy ranked first in the hierarchy of the sciences.

On Lowell's final trip to Japan (1892), his luggage included a six-inch-diameter telescope. He used this instrument to observe Saturn and other celestial bodies. Despite these observing experiences, Lowell probably decided to concentrate upon Mars only after reading about Martian life in a book by the French astronomer and popularizer of science, Camille Flammarion.

Lowell returned from Japan less than a year before October 13, 1894, a favorable time for viewing Mars from Earth. He immediately made plans to study Mars in 1894. Using his own money, Lowell decided to construct an observatory by the viewing deadline, equip it with suitable instruments, and hire a staff to work under his direction. Once Lowell made this decision, he turned away from the Orient and made astronomy his main concern in life.

The Creation of Mars

Lowell needed the assistance of experienced astronomers in choosing a site for the new observatory and hiring its staff. He turned to the Harvard College Observatory for advice. Its director, Edward C. Pickering, released his younger brother William from the Observatory staff to help Lowell plan his new effort. This is the same William Pickering who observed melting snow and snowstorms on Mars in 1892.

Lowell sent William Pickering and Andrew E. Douglass to find the proper location for his observatory. Pickering convinced Lowell that the best site for planetary observation was one where the air was relatively calm. The Earth's turbid atmosphere, he argued, created problems for earlier observers of Mars. When Pickering suggested a desert location in Arizona Territory, Lowell immediately ordered Douglass to survey sites in the region.

Eager to start construction of his observatory, Lowell chose Flagstaff from a list of Arizona sites proposed by Douglass. In later years, Lowell boasted that superior viewing conditions at Flagstaff, and his use of smaller telescopes, enabled him to see Martian canals when astronomers using larger instruments failed to find them. Lowell's critics reluctantly conceded the superiority of the Flagstaff location. They did not know that Lowell became dissatisfied with Flagstaff within a year and considered moving his observatory to Mexico.

Lowell chose the Flagstaff location on April 16, 1894. On May 22, he appeared before the Boston Scientific Society, where he stated the goals of his new observatory. Although Lowell had not yet observed the Martian canals and had not even visited Flagstaff, he was ready to present his conclusions on the subject.

Lowell told his Boston audience that the possibility of finding life on other worlds, including intelligent creatures, was not a fantastic notion. There was strong reason to believe, he continued, that important discoveries were imminent. The *canali*, for example, were obviously the work of intelligent beings.

Lowell arrived at Flagstaff six days after making these sensational claims. On the night of May 31, he finally was able to view Mars through a telescope at his hastily built observatory. Astronomers with more experience in observing Mars criticized Lowell for reaching conclusions on a controversial topic before he had made a single observation of the planet. An astronomer at the Allegheny

Observatory in Pittsburgh remarked that Lowell drew no distinction between what he saw and what he inferred.

Lowell's entrance into the canal debate repeated what he had done in Japan. Two weeks after arriving in the Orient, he claimed to understand the soul of the East. Lowell interpreted the Oriental mind for Westerners although he lacked fluency and literacy in the Japanese language, spent most of his time with Westerners in Tokyo, and avoided Japanese food. Now, as an astronomer, he announced the solution to the Martian controversy before he approached the eyepiece of a telescope. In both instances, it was a matter of hastily drawn generalizations used to buttress a theory believed to be impervious to change or criticism.

Schiaparelli's 1893 essay on the subject strongly influenced Lowell's conception of Mars. At first Lowell accepted the Italian astronomer's use of terrestrial terms to describe the physical features of Mars, his belief that Mars was an aquatic planet, and his conception of irrigation *canali*. Schiaparelli was grateful for Lowell's support, but he was uncomfortable with some of his speculations and the way he publicized them.

Once he began studying Mars, Lowell saw, sketched, and counted the canals he found and observed the process of canal doubling. After a month of astronomical work at Flagstaff, Lowell returned to Boston, leaving the observatory in the hands of its staff. In July 1894, Pickering, using a polariscope to analyze light reflected from the dark regions of Mars, discovered that there was no water in Schiaparelli's Martian seas.

Lowell accepted Pickering's findings. He incorporated them into a new vision of Mars as a planet largely covered with deserts. The dark regions of the planet observed by astronomers were areas of vegetation, not bodies of water. The melting of the polar ice caps during the warm season freed water to flow through the canal system. The flowing water irrigated the desiccated planet and brought life to its vegetation. Lowell's theory, completed after a short stay at Flagstaff, changed little over the next twenty years.

In describing the orderly arrangement of the Martian canals, Lowell compared them to trigonometric figures. Lowell's maps of the canals are simpler and more geometrical than Schiaparelli's. There are two explanations for Lowell's schematic maps. First, Lowell studied Mars using Schiaparelli's maps as his guide. Second, according to Carl Sagan, Lowell was a poor draftsman who drew polygonal blocks linked by many straight lines. Pickering and Douglass were no better at rendering details of the Martian surface than Lowell.

With his theory completed, Lowell was ready to bring it to the world. The Lowell Observatory had its own technical publication—*The Annals*—but Lowell hoped to reach a larger readership. He announced his results to newspapers, wrote articles for popular magazines, and lectured to large audiences. Lowell's media campaign reached its peak with the appearance of a two-hundred-page book entitled *Mars* (1895).

Percival Lowell offered a comprehensive theory of Mars based on Herbert Spencer's evolutionary philosophy. The unique physical conditions of the planet, Lowell declared, explained the social behavior and technology of the intelligent creatures who lived there. Lowell claimed that because Mars was smaller than the Earth, it evolved faster. Mars continued on its rapid evolutionary path and soon reached the final stages of planetary development.

Lowell believed that Mars was older than the Earth. All planets, Lowell argued, become drier as they age. At one time, the Earth had much more water than land. On Mars, land had largely replaced water, leaving the planet covered with vast desert regions of a reddish-ochre color. This color reminded Lowell of the Sahara region of northern Africa or the Painted Desert of Arizona.

Mars was dry but not without water. Lowell drew attention to Martian polar caps that melted during the warm seasons. As the polar caps retreated, a deep blue band appeared around the poles. This band was ice that melted with the rising temperature of the Martian spring and summer. Lowell dismissed the hypothesis that the polar caps were largely frozen carbon dioxide, not snow and ice.

A desert planet with water frozen in polar ice caps is an unlikely habitat for life. However, Lowell assured us that Mars had enough water to sustain life. It also had an adequate supply of air, another crucial ingredient of life. Lowell's telescopic study of the disk of Mars convinced him, if not other astronomers, that Mars had an atmosphere. Water circulated throughout the atmosphere in a vaporous form and condensed at the poles of the planet.

The freezing and melting of water at the polar caps convinced Lowell that the average temperature of Mars was comparable to the Earth's, if not higher. Hence, Martian polar caps shrink back far more drastically during the warm seasons than do the ice caps at the Earth's poles. The existence of water and air on Mars, along with its mild climate, were essential to Lowell's picture of the planet as a place of constant change. It was not static like the airless, waterless, and lifeless Moon.

Lowell first described the physical characteristics of Mars. Then he was ready to introduce life there. The Martian climate, smooth terrain, and adequate supply of water and air indicated life could thrive on the planet. If astronomers properly examined Mars through their telescopes, evidence of life would emerge.

Lowell claimed that the dark, bluish green regions of Mars turn to shades of gray and brown seasonally. The dark areas are plants that flourished with warmth and moisture and faded when the frosts of the Martian autumn arrived. The changing colors of Martian vegetation reminded Lowell of American forests seen from a distance.

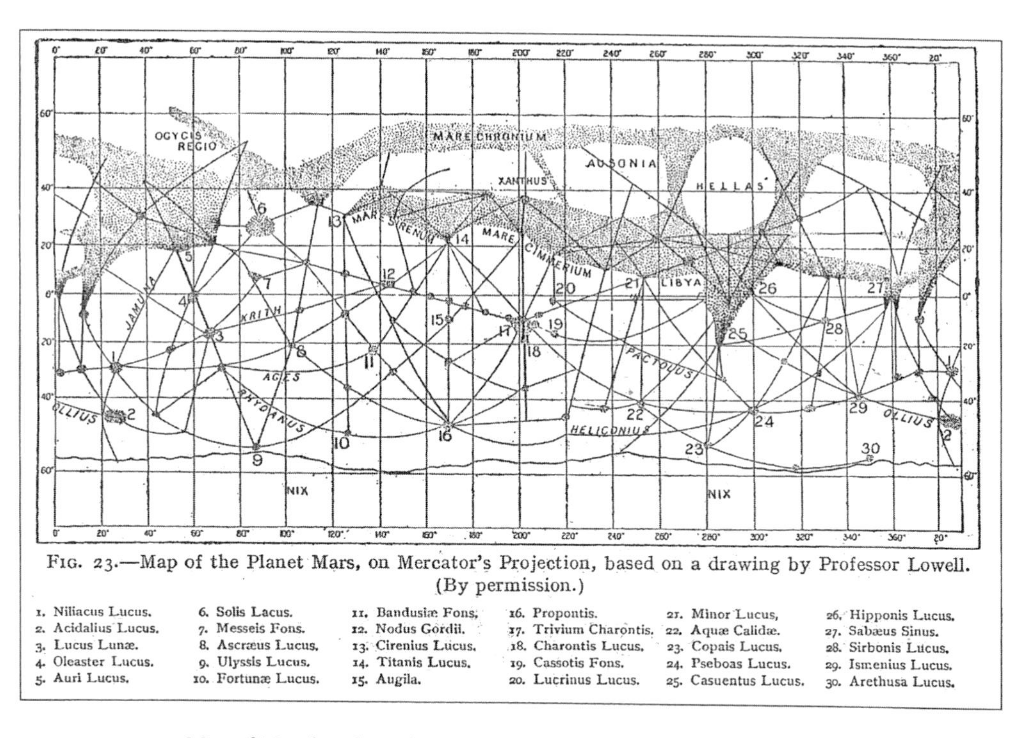

FIG. 5.1. Map of Mars based on observations made by Lowell. Note the crisp, clear lines of the canal system. (Camille Flammarion, *Astronomy for Amateurs*. London: T. Nelson, 1903.)

After establishing the existence of vegetation on Mars, Lowell turned to Schiaparelli's *canali*. He saw them in his telescope as a network of dark straight lines girdling the planet (Fig. 5.1). At the junctions of several lines, Lowell observed large circular or triangular shapes. He said the canals were as clear to him as the lines in a fine steel engraving.

Lowell argued that the lines he saw on Mars were artificial because they were ruler straight, uniformly wide, and interconnected to form a system. The regularity of the canal network so impressed Lowell that he compared it to the geometrical layout of the walkways in London's Hyde Park. Lowell was familiar with these walkways. He had studied them from a height of 5,500 feet during a hot-air balloon ascent over the park in 1908.

Lowell believed that nothing on Earth of natural origin was comparable to the mesh of lines displayed on Mars. Terrestrial streams or riverbeds did not duplicate the geometrical complexity and regularity of the Martian network, nor did cracks or ravines on the Earth's surface mimic them. The lines, unique to the red planet, were the result of special physical conditions that prevailed there.

The water supply on Mars was limited to its polar regions, yet vegetation grew at specific locations over the Martian landscape. Close study of the planet revealed that there was a temporal connection between melting polar caps and flourishing vegetation. The revival of vegetation appeared *after* the polar ice melted.

Lowell thought that there was only one way to explain the connection between melting polar caps and the appearance of vegetation. Martians diverted water to vegetation in distant desert regions during the growing season. The harsh Martian environment forced its inhabitants to engage in large-scale irrigation to preserve a dwindling water supply. The details of Martian irrigation technology might remain unknown except that the system installed on Mars was visible to Earthbound observers using telescopes.

Lowell emphasized that the dark lines observed on Mars change over time. They even disappeared entirely at some point. Significantly, after one or more lines disappeared they reappeared in the same place. How can we account for this strange phenomenon? Lowell's solution was that we do not observe changes in the canals themselves because their narrow channels were not visible from Earth. The changing dark lines were broad strips of vegetation planted along the sides of irrigation canals.

Lowell was the first to see large round or oval patches, along with dark triangular areas, situated directly on the canal network. He observed and named 186 dark round spots and calculated that each one was 120 to 150 miles in diameter. Since the spots appeared and reappeared along the dark lines, Lowell concluded that they were large oases consisting of irrigated vegetation surrounding a smaller urban population center. The oases blossomed when water from the polar caps reached them seasonally.

The only remaining puzzle was the triangularly shaped areas, which Lowell called "carets." The placement of carets at key points in the network suggested they were relay stations for the water before it entered the canals. The discovery of relay stations did not explain how water in the planet-wide canal system flowed for hundreds, if not thousands, of miles, helped by the force of gravity alone. Lowell acknowledged the problem and proposed that Martians used contrivances of some sort to move water along the far reaches of the planetary irrigation system.

Lowell noted that a good telescope could detect details of the life cycle of Martian flora, but it was useless for viewing Martian fauna. Intelligent animal life, however, had other ways of making itself known. The first action of intelligent life on any planet, explained Lowell, was the domination of nature. As the brain evolved, it progressively gained control over the natural environment.

Lowell believed that the control of nature and the imposition of geometric patterns on the landscape accompanied the evolution of human culture. Spencer's evolutionism explained the emergence of intelligence and culture. Lowell offered specific examples of this evolutionary process in action. Early farmers learned that it was more efficient to plow and plant in ever straighter rows. Old crooked footpaths were first replaced by more regular roadways, and then by precisely aligned railroad tracks. Humans destroyed forests, dug long ship canals, and constructed sprawling communication systems on Earth. Each of these activities impressed large-scale geometrical grids upon the Earth's surface.

According to Lowell's planetary timetable, intelligent beings conquered nature earlier on Mars than on Earth. Therefore, the canal network was the physical manifestation of the older, superior intelligence inhabiting Mars. The Martian mind stamped itself indelibly upon the Martian landscape. It was there for us to see and study.

Lowell warned of the danger of thinking that Martians were physically similar to humans. As he put it, the existence of extraterrestrial life did not necessarily mean "real life in trousers."[2] In the thin atmosphere of Mars, intelligent creatures might breathe through gills, not lungs, he noted. Because of Mars small size, and lower force of gravity, Martians could be several times larger in stature than humans. Such gigantic creatures would find it easy to excavate a planet-wide canal system.

Lowell's reluctance to supply details about his Martians disappeared when he considered their intelligence. Mars was much older than the Earth, and hence, Martian life had advanced far beyond human intelligence. The Martian canal system implied a mind greater than that "which presides over the various departments of our own public works."[3] Unlike their inferiors on Earth, Martians had risen above petty party politics and arbitrary national boundaries to govern on a planet-wide basis.

Lowell concluded his tribute to the Martian intellect by speculating that Martian inventions surpassed our wildest technological dreams because they "are in advance of, not behind us, in the journey of life." Consider the telephone and moving-picture machine, both recent inventions in Lowell's day. Lowell claimed that on Mars these devices were "preserved with veneration in museums as relics of the clumsy contrivances of the simple childhood of the race."[4]

Lowell was not the first to claim intellectual and technological superiority for extraterrestrial life, but he was the most influential modern writer to reach that conclusion. After Lowell, the technological superiority of extraterrestrial beings became a fundamental assumption about intelligent life on other worlds.

Lowell's speculations about the Martian intellect did not detract from his telescopic observation of the planet. He and his assistants recorded 183 canals during the first year at Flagstaff. Schiaparelli had already charted 67 of these canals. By the end of his astronomical career, Lowell had drawn more than 700 canals on his maps of Mars and determined that at least 400 of them were over 2,000 miles long. Whatever his faults and biases, Lowell was a committed, tireless observer. He advanced our interest in the features of Mars, if not our deeper understanding of them.

The Eye of the Beholder

Lowell was not content to view Mars through a telescope. He wanted to photograph its canals. A photograph of the planet with the canals clearly displayed would provide the objective proof he needed to silence critics and win scientific approval for his views. Personal bias, bad eyesight, or poor draftsmanship can distort the human observer's interpretation of Mars. A camera, on the other hand, was a neutral recorder of reality.

Unfortunately, at the end of the nineteenth century, it was difficult to join camera to telescope and obtain satisfactory images of planetary surfaces. Solar and stellar photography had advanced during the nineteenth century but not planetary photography. Photographing planets posed special problems because the definition of details was more important than in stellar photography, whose goal was the capture of the illumination. As late as the early 1970s, planetary observers relied upon observation and sketching to obtain the finer details of a planet. This situation changed when the first spacecraft circled a planet, collecting images of its surface.

In the early 1900s, the Lowell Observatory staff changed the course of astronomical photography by developing the first technical apparatus and procedures that made it possible to photograph a planet. Carl O. Lampland, working at Flagstaff, did pioneering research in planetary photography. Lampland made seven hundred photographs of Mars when it was in a prime viewing position in 1905. Lowell announced that not only had Lampland captured Martian canals on his photographic plates, but he had also photographed the first snowfall of the season near the Martian north pole. News of these accomplishments appeared in the popular and scientific press of Europe and America.

The scores of articles describing Lampland's groundbreaking work did not include photographs of the canals. Lampland had succeeded in obtaining photographic images of Mars, but they were very small pictures—about one-quarter inch in diameter. It was impossible to enlarge these tiny delicate images for transfer to the printed page of a journal or newspaper without first retouching them. For that reason, Lowell could not print Lampland's photographs in his popular books on Mars.

Because enlarged printed photographs of the canals were not available, astronomers either accepted Lowell's evaluation of the images or inspected Lampland's original plates personally. Everyone agreed that Lampland had succeeded in photographing the face of the planet. They disagreed about the existence of Martian canals. Some claimed they saw canals, others did not.

There were even skeptics among the astronomers who saw canals on Lampland's photographic plates. They reasoned that the optics of the camera, like the eye of the human observer, fused small discrete landscape features into straight lines. Lowell soon learned that photographs did not capture reality. They were as open to criticism and interpretation as hand-drawn sketches and printed maps of Mars.

Once Lowell formalized his thinking about Martian canals and presented it as a full-blown scientific theory, he resisted any efforts to change it. Contrary observational evidence, and growing criticism of his views, could not persuade him to abandon or change any part of the general theory he formulated in the summer of 1894. Lowell's adamant defense of his conception of Mars lasted until his death in 1916.

When discussing the critical response to Lowell's theory, it is important to recall the handicaps astronomers faced when they observed Mars. Astronomers were puzzling over details in the physical features of a planet located at least 35 million miles away, and whose average size in the sky was 1/100 the diameter of the full Moon. Observation of Mars took place at the limits of visual perception using instruments operating at the edge of their resolution and magnifying powers.

Critics organized their response to Lowell's canal theory along three lines: the physical conditions of Mars, the illusionistic nature of the observed linear markings, and the inadequacies of Lowell's method of scientific inquiry. These criticisms were old. Schiaparelli had grappled with them earlier. However, the accumulation of new data, and the determination to resolve the Martian canal question, motivated astronomers to reassess Lowell's conception of Mars.

Critics questioned every aspect of Lowell's depiction of the Martian physical environment. They said that the Martian atmosphere was too thin to sustain life and the temperature too low to allow for the extensive melting of polar ice. Finally, they calculated that the Martian water supply was insufficient to fill the canals and irrigate desert lands.

A growing number of astronomers argued that the dark lines Lowell saw on Mars were optical illusions created by the observation process. These critics implied that Martian canals belonged to the field of perceptual psychology, not astronomy. Questions about *how* astronomers perceived objects became important when observation took place at the limits of detection.

Lowell viewed, discovered, and drew more canals than any other astronomer. He attributed his success to his acute eyesight and to the still air at Flagstaff. Yet other Flagstaff observers, using the same telescope, on the same night, failed to see the same set of canals. American and European astronomers met similar problems when they tried to duplicate Lowell's observations using their instruments.

As canals appeared, disappeared, and doubled, astronomers began to realize that they were experiencing optical illusions, not viewing the physical features of Mars. Lowell's ambitious plans to photograph Martian canals failed. His public announcement that he had observed two new canals under construction on Mars within the past few months was met with skepticism.

Critics often attacked his approach to science. They accused him of bringing preconceived theories to his work and of obstinately refusing to change his mind when confronted with contrary evidence. Greater scientists than Lowell have exhibited similar faults, but Lowell did show an extreme rigidity of thought.

Lowell's claim that intelligent beings on an arid planet must construct canals rightly drew criticism. This unsupported claim rested upon a simplistic environmental determinism. If Mars was a desert planet, then its inhabitants had to engage in irrigation agriculture. The mode of agriculture, in turn, determined the nature of Martian civilization.

Historians of astronomy generally agree that the turning point in the canal controversy came in 1909 when the distinguished Greek-born astronomer Eugène M. Antoniadi (1870–1944) observed Mars with the thirty-three-inch-diameter refractor telescope at Meudon, France. Antoniadi, a skilled observer, made the best use of Europe's largest telescope to study Mars when viewing conditions were very good.

After several months of observation, Antoniadi concluded that the canal networks of Schiaparelli and Lowell were an optical illusion. Instead of canals, Antoniadi found "myriads of marbled and chequered objective fields, which no artist could ever think of drawing."[5] Antoniadi's assessment carried extra weight because he was an excellent draftsman (Fig. 5.2).

Antoniadi's powerful telescope resolved the dark linear markings into diffuse streaks and borders of shaded areas similar to those depicted on Nathaniel Greene's map of 1877 (Fig. 5.3). In the United States, George Ellery Hale confirmed Antoniadi's findings using a sixty-inch-diameter reflector. The American astronomer found "small irregular dark regions"[6] but no canals.

Lowell responded vehemently to these new studies of Mars. He wrote to the editors of *Scientific American* that opposition to canals came "solely from those

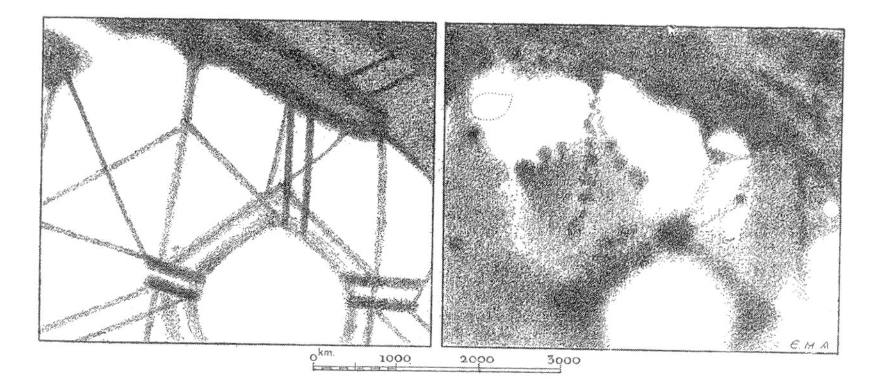

FIG. 5.2. Left panel displays Schiaparelli's observations (1877–1890) of single and double canals in a region of Mars. Right panel shows the same region, filled with shaded masses but no canals, observed by Antoniadi (1911, 1924, 1926). (E. M. Antoniadi, *La Planète Mars*. Paris, 1930. Hermann éditeurs des sciences et des arts.)

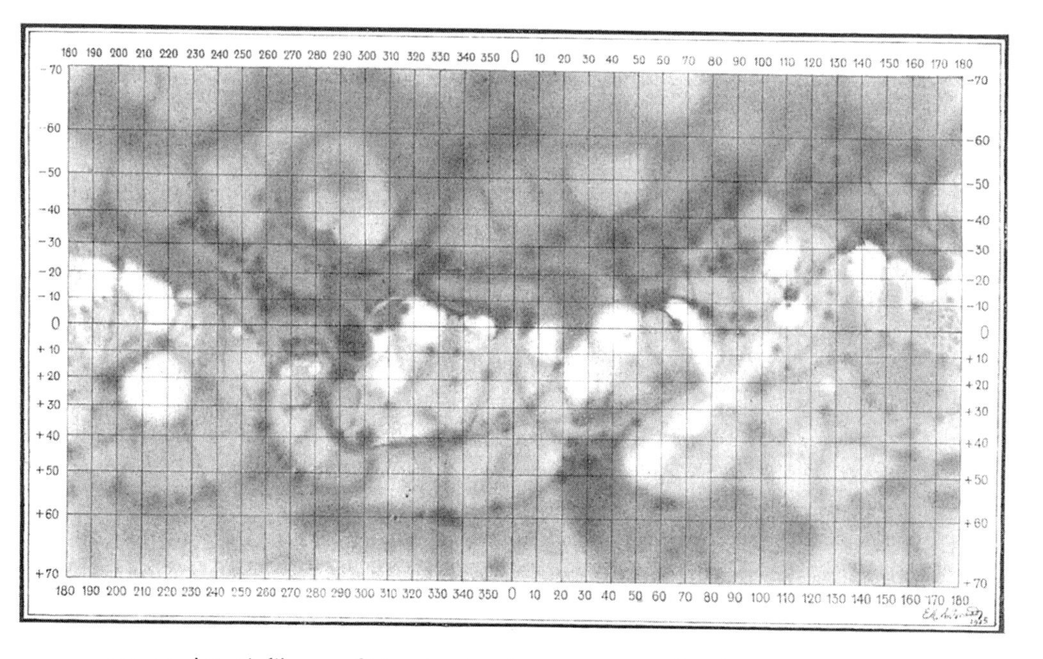

FIG. 5.3. Antoniadi's map of Mars showing the average state of its changing surface up to 1929. (E. M. Antoniadi, *La Planète Mars*. Paris, 1930. Hermann éditeurs des sciences et des arts.)

who without experience find it hard to believe or from lack of suitable conditions find it impossible to see."[7] Lowell's modern biographer, W. G. Hoyt, commented that Lowell's response to the editors was both self-serving and slanderous.

Lowell was as certain of the absolute validity of his Martian theory as he was of his discovery of the essence of the Oriental soul. In both instances, there was no turning back to modify or rethink his first conclusions. Shortly before his death, Lowell offered a final assessment of his theory:

> Since the theory of intelligent life on the planet was first enunciated 21 years ago, every new fact discovered has been found to be accordant with it. Not a single thing has been detected which it does not explain. This is really a remarkable record for a theory.[8]

Lowell's evaluation of his life's work is delusional. By 1916 his theory was under strong attack. His canals, oases, and vegetation soon disappeared in a haze of optical illusions. Large-diameter telescopes, whose use Lowell belittled, showed vast indistinct areas of varying shades on a Mars with no canals. Although a part of the public remained faithful to Lowell's conception of Mars, a dwindling minority of astronomers accepted the existence of canals.

The final resolution of the nature of the Martian landscape came in the space age. American spacecraft of the Mariner missions passed over the planet, recording its surface features (1965–1971). Mariner cameras disclosed an arid landscape filled with craters, ravines, and extinct volcanoes. There were no signs of artificial constructions or vegetation.

Lowell's America

America experienced some important changes during the years Lowell was observing and interpreting Mars (1894–1916). These changes helped to shape his conception of Martian life and culture.

European and American inventors and entrepreneurs introduced automobiles, airplanes, electric lighting, telephones, phonographs, motion pictures, and cheap consumer goods. These and related developments led many contemporary thinkers to declare that theirs was an age of unrivaled technological and social progress.

Not all Americans in the nineteenth century were certain they were living on the brink of utopia. Critics opposed to the prevailing conception of material progress emerged in the 1890s. Among them were Americans influenced by European intellectuals caught up in the melancholy mood of the *fin-de-siècle* (end of the century). According to these critics, the recent past showed unmistakable signs of decline and degeneration, not progress. One source of *fin-de-siècle* gloom in Europe was the belief that a war between the major European powers was inevitable.

Although Percival Lowell did not share the pessimism of these critics, *fin-de-siècle* thought influenced his theory of Mars. Lowell, who had an excellent command of the French language, became a close friend of the French astronomer Camille Flammarion. In 1893 Flammarion published *La fin du monde*, a book translated and read around the world. Its English title was *Omega: The Last Days of the World*. Flammarion's novel melded *fin-de-siècle* with *la fin du monde* to fuel a popular notion that the history of the human race was about to end (Fig. 5.4).

Flammarion takes his readers ten million years into the future. Humans have abolished war and established a utopian society by subjugating nature. At this point, the physical conditions that make terrestrial life possible begin to change. The internal heat of the Earth dissipates and its water supply disappears, leaving a cold desert planet. The environmental condition of the dying Earth, Flammarion notes, is similar to that of present-day Mars.

The *fin-de-siècle* melancholy that colored Flammarion's writings influenced Lowell's Mars. Lowell's Martians create a utopia based upon pacifism, universal

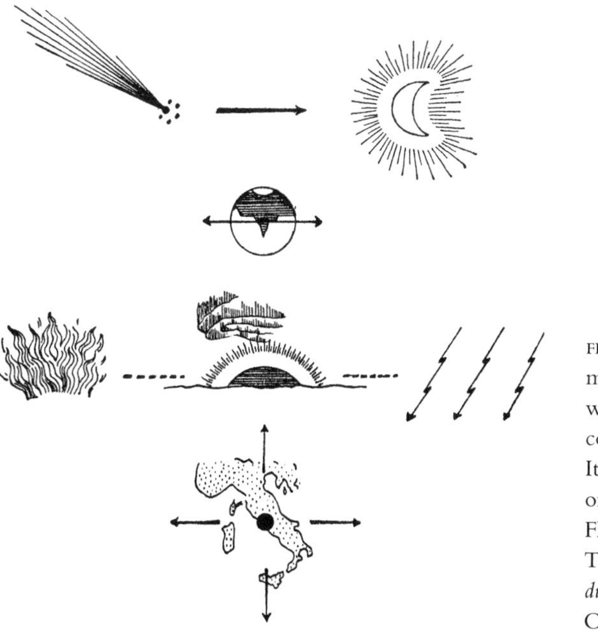

FIG. 5.4. Fictional message from Martians warning Earth that a comet will soon strike Italy (lower portion of message). (Camille Flammarion, *Omega*. Translation of *La fin du monde*. New York: Cosmopolitan, 1894.)

government, and advanced technology. However, it is a utopia born of desperation, not enlightened thought. Mars has reached the final stages of its evolution as a planet. The survival of Martian life depends upon the conservation of the planet's scant water supply in an efficient irrigation system. If Martians do not work together and innovate, they will perish. As Lowell saw it, terror was the greatest spur to the construction of the Martian canals.

Lowell refused to offer humanity any comfort because the Earth happened to be younger than Mars. This was a false hope because the two planets were simply at different stages of deterioration. Large desert regions had already begun to appear in Africa, Asia, and North America. Humans had no more chance of halting the drying up of the Earth than Martians of reversing the desiccation of their world.

Traditional Christian thought customarily allotted humans a special place in the scheme of things. Lowell did not follow its example. The existence of extraterrestrial life, he said, displaces humans from a unique or central position in the solar system. Astronomy has shown that we are a mere detail in the overall evolution of life in the universe. Humans are not the epitome of evolutionary development. They merely typify what is going on elsewhere in space. Lowell concluded that humans should not expect to find replicas of themselves among the forms of extraterrestrial life.

Lowell offered a bleak assessment of the future of Mars and the Earth because he believed they were destined to become desert planets. Deserts, and the process that Lowell called *desertism*, were of great interest to Americans living at the turn of the century. Lowell shared that interest by transferring the arid lands of the American West to Mars and using environmental determinism to account for the distinctive features of Martian civilization.

Lowell located his observatory in the mountainous desert of the Southwest, and he traveled to the deserts of North Africa and Mexico seeking optimum viewing locations. Fascinated by the desert landscape, Lowell explored the country surrounding Flagstaff studying its climate, geological formations, flora, and fauna. The information he gathered in the Arizona desert helped him to interpret life in the arid environment of Mars.

Lowell had not studied the geological and biological sciences. Therefore, he sought the advice of an old friend, zoologist Edward S. Morse (1838–1925). He was the scientist whose 1882 lectures on Japanese culture inspired Lowell to leave for the Far East. Morse suggested readings that helped Lowell to interpret the geology and natural history of the deserts of the American West. The work of C. Hart Merriam, who had surveyed the flora and fauna of the desert region near Flagstaff, was particularly useful to Lowell's investigation of extraterrestrial life.

From his reading of Merriam, Lowell learned how living organisms can inhabit a wide range of environmental niches. Merriam's observation of the fauna of the San Francisco mountain range (north of Flagstaff) showed that large animals are able to live in high altitudes, where the atmospheric pressure is low. His findings, collected within miles of the Flagstaff observatory, supported Lowell's claim that water was more critical for life on Mars than atmospheric pressure or temperature.

Lowell's study of the geological literature led him to the mistaken conclusion that the oceans of the Earth were receding and its land masses advancing. This explained why American cities found it necessary to go greater distances to tap streams and springs for urban water supply. So it must have been on Mars, said Lowell. The first hint of the problems that were to doom the red planet was the disappearance of local water sources for its population centers.

Lowell insisted that his readers have a proper understanding of deserts. Deserts are the result of the dynamic process of desertism, in which land is constantly encroaching upon water. This planet-wide process will not end until the Earth is as dead as the Moon. Lowell suggested that Americans visit the Petrified Forest of Arizona, declared a national monument in 1906, to see how deserts overwhelmed the trees of once great forests.

Lowell's examination of the deserts of Arizona and Mars should be seen in the wider context of the history of the American West. The period of public interest in Martian deserts (1877–1916) coincided with a growing national concern for arid lands in the American West.

As the immense scope of the irrigation problem in the American West became clear, a call arose for state and federal assistance. William E. Smythe was a driving force behind the newly organized national movement for irrigation. He was the founder of the journal *Irrigation Age* (1891) and secretary of the first national Irrigation Congress. In 1900 Smythe published *Conquest of Arid America* in which he described the reclamation of the desert and its conversion into fruitful irrigated farmland. A coalition of political, business, and public interest groups campaigned to make federal funds available for the construction of large storage reservoirs and their accompanying irrigation systems (Fig. 5.5).

In 1907 Lowell's friend Edward S. Morse published a book in which he discussed irrigation networks on Mars and the arid plains of the American West. If it were possible, he wrote, for a Martian to study the Earth through a telescope, "he would undoubtedly correlate the irrigating regions of Arizona as similar in nature to his own canals."[9] Morse verified his claim by observing Arizona's canal systems from nearby mountain tops (Fig. 5.6).

The era of intense interest in the irrigation of Mars and the Earth coincided with the construction of very large ship canals around the world. The Suez Canal

FIG. 5.5. Utopian view of the impact of irrigation upon the arid American West. (United States Bureau of Reclamation, *Reclamation Record*, 6. Washington, D.C.: Government Printing Office, 1915.)

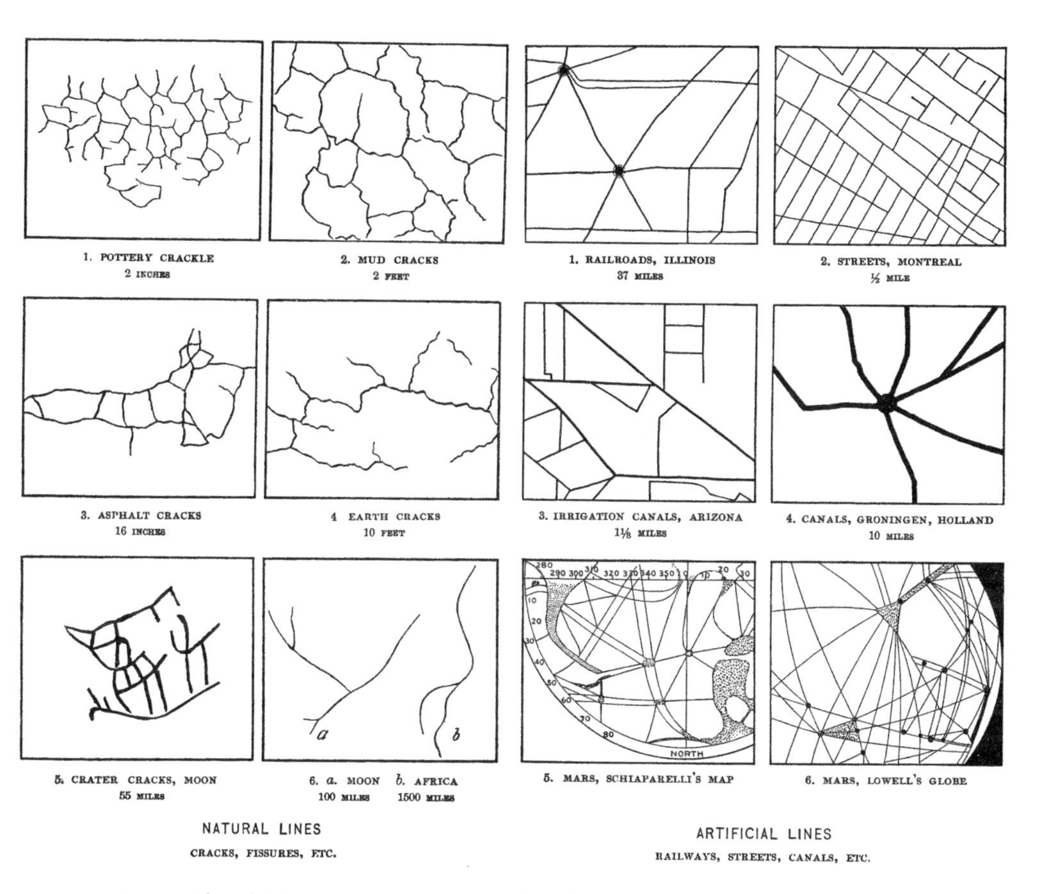

FIG. 5.6. Edward S. Morse compared natural lines (left) with artificial ones (right) to prove that the lines observed by Lowell on Mars were artificial. (Edward S. Morse, *Mars and Its Mystery*. Boston: Little, Brown, and Company, 1907.)

(1869) began the canal-building spree and the Panama Canal ended it. The canal across the Isthmus of Panama, built by American engineers (1880–1914), was hailed as one of the world's great technological projects.

Between the openings of the Suez and Panama Canals, engineers constructed a number of other canals. They built the Corinth Canal (1881–1893) in Greece; the Kiel (1887–1895) and the Oder-Spree (1887–1897) Canals in Germany; and the Manchester ship canal (1887–1894) in England. At the time Lowell was writing his book *Mars as the Abode of Life*, engineers were digging the Cape Cod Canal in his native state of Massachusetts (1909–1914).

Lowell declared that no observer of Martian canals confused them with the Suez or Panama waterways. Martians built irrigation canals, not ship canals for the transit of commercial vessels. Nevertheless, the construction of canals captured the imagination of Americans as proof of the superior engineering abilities of two great civilizations, one terrestrial, the other extraterrestrial. Specific references to terrestrial ship canals appear often in the debate over the canals on Mars.

The respected editor of the *Boston Transcript*, Edward H. Clement, wrote a long poem entitled "The Gospel from Mars" (1907). In this work, he urged humans to follow the example of Martians by building canals and rejecting war:

> The conquering heroes of the world today
> No more are butchers, but the engineers;
> Construction, not destruction, is the word.
> The era of the Canals begins on Earth.[10]

Clement's poem joined the fear of a coming great war to the current enthusiasm for building canals. It also contained a broader political message.

When Clement sent his poem to Lowell, he added a humorous aside. Since the Martians have erased national and class boundaries, he wrote, Mars truly displays the red banner of socialism as it travels through the heavens. A decade earlier, Schiaparelli maintained that a socialistic regime built and serviced the huge Martian irrigation system.

Lowell did not believe that socialists governed Mars. He was a political and social conservative whose considerable private wealth came from family holdings in Boston and in the eponymous Lowell, Massachusetts. Lowell, of course, interpreted Martian society according to Herbert Spencer's conservative social philosophy. The Martians were at their peak, he believed, because a rigorous selection process had eliminated those less fit to live. Given the conditions on Mars, Lowell wrote, only the fittest have survived. In addition to socialism, Lowell opposed labor unions, woman's suffrage, unrestricted immigration, statehood for the Arizona Territory, the abolition of the death penalty, and prohibition (because it curtailed citizens' rights).

Lowell was a lifelong foe of socialism who never lost his zeal for political discussions. In later years, he was as likely to warn his audience about the dangers of socialism as he was to lecture on astronomy. In a 1911 lecture delivered at Kingman, Arizona Territory, Lowell combined the astronomical with the political. Entitled "Two Stars," this talk was presented to win over "socialist miners," so named by Lowell, to his political point of view.

Lowell told the miners that Mars was a benevolent oligarchy ruled by an elite of technical experts. Martian technocrats assigned citizens their social roles and tolerated no dissent. The lack of water on Mars, Lowell warned, forced Martians to cooperate or perish.

Mars had a lesson for the Arizona socialists. Human society would remain weak and inefficient until all citizens learned to pull together. Pulling together meant getting rid of popular political leaders and healing the breach between capital and labor. It also meant that the average man, who lacked the wisdom and foresight to act on his own, should hand over political power to his intellectual superiors.

Lowell's political ideas were internally contradictory. The existence of a world-wide irrigation system called for the eradication of national boundaries. This, as Lowell acknowledged, led to government control of the planet's economy. Centralized government control of the economy, along with rational, social planning, are sure signs of socialism, not capitalism.

The Martians ended wars centuries ago, but not because of deliberate social policy. Lowell reported that after a number of inhabitants of a planet are killed in wars, the living find it advantageous to cooperate for the common good. Warfare is a survival from an earlier, savage stage of history. Lowell's remarks about peaceful Martians were directed to his contemporaries who feared the coming of a disastrous war.

• • •

It is difficult to evaluate Percival Lowell as an astronomer. He founded a major astronomical observatory, which still operates today, and established its initial scientific agenda. Few can dispute or match that accomplishment. On the other hand, he used the observatory and its staff to advance his pet theories about intelligent Martian life. Lowell did this in a dogmatic fashion that alienated the scientific community he sought to convert to his viewpoint. Nevertheless, Lowell's conception of Mars lasted long after his death and beyond the confines of his Observatory.

During the two decades of his scientific career, Lowell's Mars and his America gradually blended to form a single landscape. Lowell drew upon *fin-de-siècle* pessimism, environmental determinism, water-control engineering, technocratic ideals, and Herbert Spencer's conservative social theories and evolutionary philosophy. He used these to interpret the fleeting images of Mars he observed at Flagstaff.

It is easier to identify the contemporary roots of Lowell's conception of an extraterrestrial civilization than it is in the case of some other figures. Lowell did not hesitate to state his theories in bold terms. He was confident he understood Mars and Martian life better than any other person in the world.

CHAPTER SIX

Mars Unveiled

> The Mariner 4 photographs were blurry and imprecise. . . . Yet the images were clear enough to spark a worldwide controversy among astronomers. As the first pictures were retrieved from the shaded dots and converted to photographs, one could almost hear the theories and long held beliefs come crashing to the laboratory floors like the shattering of so many clay pots. Mars was not what scientists were convinced it would be.
>
> —Jay Barbree and Martin Caidan with Susan Wright, *Destination Mars*, 1997

Lowell's Legacy

Percival Lowell died of a stroke on November 12, 1916. At his death, his career was at a low ebb, and prospects were poor that his scientific work would survive. Scientists in Europe and America attacked his conception of Martian canals and noted that research conducted at his observatory was not at the forefront of twentieth-century astronomy. Meanwhile, Lowell's widow contested his financial legacy to the Lowell Observatory at Flagstaff and won, leaving it inadequately funded.

Astronomer Clyde Tombaugh, who joined the Observatory after Lowell's death, said in 1980 that controversies over the sighting of Martian canals led professional astronomers to ignore the Observatory's work. His assessment of Lowell's negative influence received support from Carl Sagan, a well-known planetary astronomer and proponent of extraterrestrial life. Sagan claimed that Lowell bore partial responsibility for the shortage of trained planetary astronomers in the 1960s. Young astronomers, said Sagan, entered stellar astronomy, where the opportunities were greater and where they could avoid fruitless debates over Martian canals. Although accurate to a degree, neither of these evaluations is the final word on Lowell's scientific legacy.

Lowell's unyielding defense of his Martian theories, and the sensational public claims he made about his discoveries, hurt his standing among scientists. On the positive side, Lowell was an advocate of solar system astronomy at a time when stellar astronomy dominated astronomical research.

Lowell established a private research observatory that remains a center for the study of planets within our solar system. He directed his able staff to do pioneering experiments in planetary photography, engage in spectroscopic studies that uncovered evidence of an expanding universe, and search for an unknown body (Planet X) beyond Neptune.

Lowell devoted a great deal of time, effort, and money in the last years of his life searching for the elusive Planet X. After calculating the planet's orbit and failing to find it, Lowell abandoned his quest in 1915. During the 1920s, the Lowell Observatory staff renewed efforts to locate the trans-Neptunian planet. In 1930 Clyde Tombaugh discovered Pluto, an event that brought lasting fame to him and the Observatory. Although the planet did not travel in the orbit predicted by Lowell, its discovery was part of a research project he initiated.

Lowell's popularity may have waned within the scientific community, but it remained high with the mass audience he cultivated in his lectures and books. His influence in the public arena grew when several early science fiction writers set their stories in the Mars he portrayed. Readers of these tales were probably not familiar with Lowell's astronomical work, yet they came to view Mars as a place of deserts, canals, and a doomed race of intelligent beings.

Three writers, the Englishman H. G. Wells (1866–1946), the German Kurd Lasswitz (1848–1910), and the American Edgar Rice Burroughs (1875–1950), produced enormously popular novels set in a Lowellian Mars. Wells cleverly combined the story of a dying Mars with the theme of a coming great war in his novel *The War of the Worlds* (1897). Wells's Martians leave their decaying planet to conquer the Earth with the help of advanced weapon and transportation technology. Their invasion fails when they succumb to bacterial infections harmless to humans. The American film director Orson Welles broadcast a radio version of *The War of the Worlds* in 1938. The panic created among American listeners by this radio drama testifies to the widespread acceptance of Lowell's version of the red planet in the 1930s.

Lasswitz's *Auf Zwei Planeten* (*On Two Planets*), first published in 1897, tells of a more peaceful invasion by Martians who come to Earth in search of raw materials unavailable on their aging and densely populated planet. The Martian invasion leads first to conflict and then to the establishment of a utopian world state on Earth. The scientific accomplishments of Lasswitz's Martians, which surpass anything on Earth, are less important than the ethical instructions they impart to humans. *Auf Zwei Planeten* was the most popular science fiction novel on the Continent at the turn of the century.

Edgar Rice Burroughs was the foremost American literary interpreter of Lowell's Mars. He created Tarzan of the apes and wrote eleven novels set on Mars. Burroughs's Mars, known to its inhabitants as Barsoom, is the scene for exotic travel and adventure tales featuring the heroic American explorer John Carter. Burroughs's books enjoyed great popularity in the United States.

The Mars of these and other science fiction writers exerted a powerful influence on popular attitudes toward the American space program and on key figures in space science. Pioneers of space flight, as well as the first astronomers to search for extraterrestrial intelligence with radio telescopes, acknowledge their debt to the Mars of popular culture.

The German-American rocket builder Wernher von Braun recorded his obligation to Lasswitz in an introduction he wrote to the first English translation of *Auf Zwei Planeten*. Physicist Freeman Dyson and astrophysicist Philip Morrison, two well-known scientists engaged in the search for extraterrestrial intelligence, claimed that reading H. G. Wells at an early age stimulated their interest in alien life.

When Carl Sagan was a young boy of six or seven, he read Burroughs's John Carter stories and never forgot the thrill they aroused in him. Barsoom remained a magical place for Sagan into his adult life. He fondly remembered it as a "world of ruined cities, planet-girdling canals, immense pumping stations—a feudal technological society."[1] For twenty years, a map of Barsoom hung in the hallway outside of Sagan's Cornell University office. Science fiction remained a powerful influence on Sagan's scientific thought throughout his career.

The power of the image of Mars found in popular culture was exceptional and even affected scientists. Late in 1971, a group of scientists and writers met publicly to discuss the implications of the Mariner 9 spacecraft sent to record close-up pictures of the surface of Mars. The discussants included Carl Sagan, science fiction writers Ray Bradbury and Arthur C. Clarke, journalist Walter Sullivan, and Dr. Bruce Murray. Murray, a California Institute of Technology professor of planetary science whose specialty was the geology of Mars, later served as an administrator for various NASA-sponsored SETI projects. On this occasion, Murray argued that Mars had such a powerful grip on human emotions and thoughts that it distorted scientific opinion regarding life on the planet. He warned:

> My own personal view is that we are all so captive to Edgar Rice Burroughs and Lowell that the observations are going to have to beat us over the head and tell us the answer in spite of ourselves.[2]

Murray was referring to working scientists, not to science fiction fans. He worried that scientists would interpret Mariner 9 images within the old Lowellian framework. Murray's only hope was that unambiguous and compelling data from the Mariner spacecraft would force scientists "to recognize the real Mars." Such was the strength of Lowell's legacy a half century after his death.

• • •

Early twentieth-century astronomers, physicists, chemists, and biologists preserved features of Lowell's Mars. They studied the chemical constitution and pressure of the Martian atmosphere and investigated the existence of water and vegetation on the planet. Scientists who rejected Martian irrigation canals, along with their alleged builders, retained other aspects of Lowell's conception of the planet. They believed that Mars was an arid but essentially Earthlike body with polar ice caps. Seasonal melting of the polar ice released a flow of water that triggered the spread of plant life across the planet's surface.

A majority of prominent astronomers polled by the *New York Times* in 1928 agreed that plants and perhaps simple animal forms existed on Mars. Interest in Martian vegetation increased in 1947 when astronomer Gerard P. Kuiper (1905–1973) discovered carbon dioxide in the Martian atmosphere. The presence of carbon dioxide meant that plants containing chlorophyll might thrive on the planet. Kuiper noted that his findings did not rule out the possibility of lichens or mosses on Mars.

In 1953 Hubertus Strughold, a pioneer in aviation medicine, raised the possibility of finding microbial life on Mars. Strughold proposed the construction of a chamber that simulated the physical conditions of Mars in the laboratory. Biologists could then test a variety of terrestrial organisms in the artificial Martian atmosphere of the chamber.

The hypothetical flora and fauna of Mars had degenerated considerably from the heyday of Lowell when a race of superior technocrats governed the planet and built a vast canal system. By the middle of the twentieth century, lichens and microorganisms were envisioned as representative Martian life forms. Even at that low level, the possibility of life on Mars generated sufficient scientific and popular interest to inspire early space efforts.

When the Soviet Union launched the first artificial satellite (Sputnik) in 1957, Mars was viewed as a planet with harsh physical conditions that might contain life of some sort. The United States government's decision to explore the Moon in the late 1960s meant there was less money available for manned planetary exploration. Unmanned voyages to the nearest planets, however, were technologically feasible and required smaller budgets. Mars was a popular destination for these voyages.

The race between the United States and the Soviet Union to reach Mars began in 1960 when the U.S.S.R. attempted to launch a spacecraft intended to fly near the planet. The Soviet craft never reached Earth orbit. The Americans won the race to Mars in 1965. Mariner 4 spacecraft made a close sweep over Mars, transmitting twenty-two images of the planet to NASA's Jet Propulsion Laboratory in California. Before the launch of Mariner 4, the official *NASA Sourcebook on Space Sciences* concluded that "most astronomers would probably agree that there are apparently linear markings . . . of considerable length on the surface of Mars."[3] The markings, of course, were those observed and interpreted by Schiaparelli and Lowell.

The first close-up pictures of Mars surprised NASA scientists. There were no channels, artificial or natural, on the planet, nor were there any signs of life. Instead, large craters covered the Martian landscape. A century of close telescopic observation had not disclosed the existence of craters on the planet. Lowell had argued that the cratered Moon was sterile but that Mars and Earth were sister planets with similar landscapes. Mariner 4 recorded the features of the new, post-Lowellian Mars (Fig. 6.1). Its discoveries were presented in a *U.S. News and World Report* article that read: "Mars is dead. There are no cities, oceans, mountains, or even continents visible on Mars."[4]

Carl Sagan, ever the optimist when extraterrestrial life was the issue, pointed out that Mariner recorded the Martian surface from an altitude of 6,000 miles. The Earth photographed at that height would also appear lifeless, he added. Sagan offered a personal interpretation of the Mariner 4 images for the readers of the 1967 *National Geographic* magazine. In an article illustrated with hypothetical Martian flora and fauna, he claimed that it was possible for organisms to adapt to the rigorous environment of Mars.

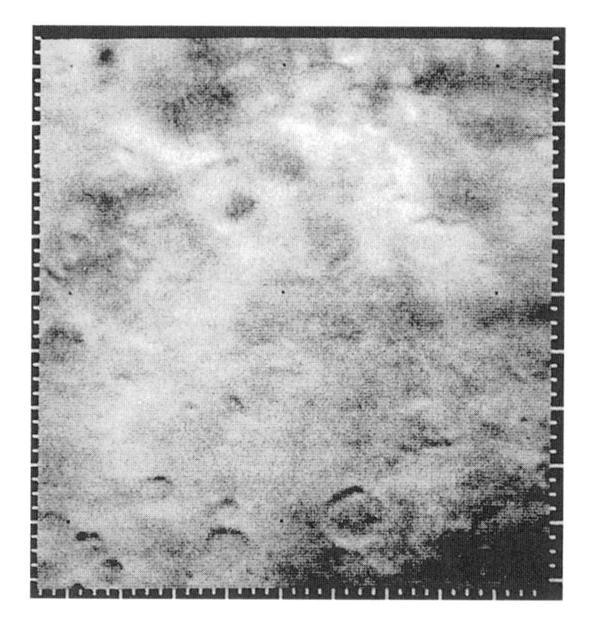

FIG. 6.1. First picture showing craters on Mars. Returned by Mariner 4 spacecraft in 1965. (Courtesy of National Space Science Data Center.)

Hopes Dashed

Mariner spacecraft sent to Mars between 1964 and 1969 revealed a Martian environment that was more hostile to life than Lowell's critics ever imagined. Mars was a dry, frigid planet with polar caps of frozen carbon dioxide along with some ice. The thin Martian atmosphere consisted largely of carbon dioxide. Atmospheric pressure on Mars was less than 7 millibars compared to 1,000 millibars at the Earth's surface. The low atmospheric pressure and temperature of Mars meant that water could exist only as a vapor or a solid.

In 1971 a NASA spacecraft traveled to Mars to record its surface in detail and construct the first comprehensive map of its physical features. Mariner 9, placed in orbit around Mars, carried a television camera capable of relaying high-resolution images (Fig. 6.2). Carl Sagan and Robert Fox carefully analyzed the newly gathered images. They searched for a correlation between the detailed maps of Mars drawn by Schiaparelli and Lowell and the physical features displayed in Mariner 9's excellent pictures.

The two astronomers found no correlation between the old maps and Mariner 9's images. A small number of canals may correspond, they noted, to identifiable topographical features, but the vast majority of canals "appear to have no relation to the real Martian surface" (Fig. 6.3).[5] Sagan and Fox concluded

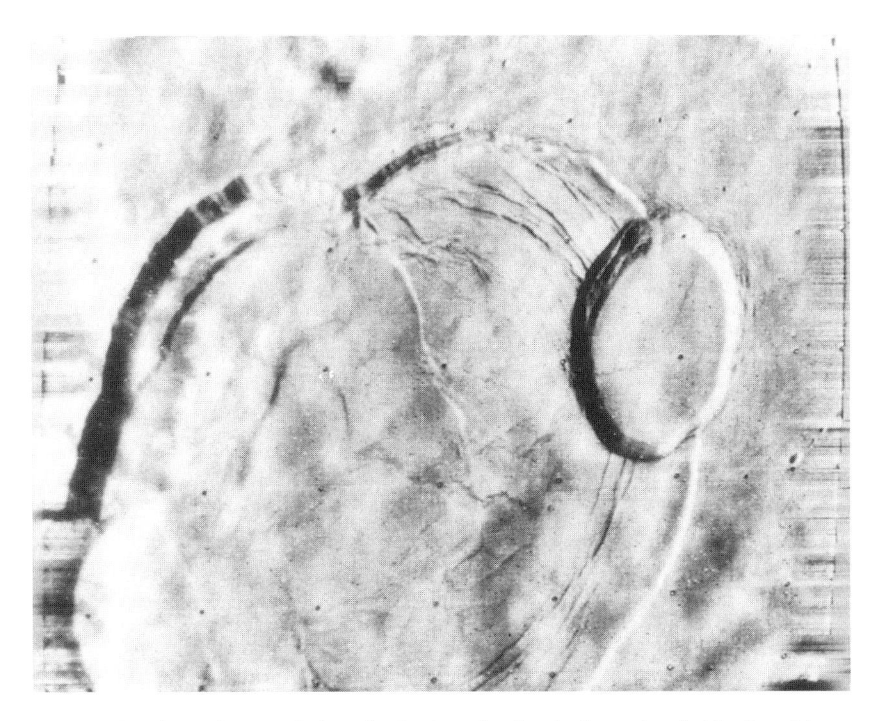

FIG. 6.2. Mariner 9 image of a large basin-shaped volcanic depression (caldera) on Mars. (Courtesy of National Space Science Center: Principal Investigator, the late Dr. Harold Masursky.)

that the classic maps of Martian canals resulted from faulty attempts to sketch the planet while observing it under difficult viewing conditions.

Some of the visual evidence collected by orbiting spacecraft supported Lowell's interpretation of Mars. The Mariner 9 photographs showed channels resembling dry river beds. However, they also revealed large inactive volcanoes that Lowell had neither seen nor assumed. Stream beds appeared carved into the surface by water flowing long ago, perhaps billions of years ago. Ancient water erosion and volcanic activity suggested that Mars was once a more dynamic place than it is today.

Signs that water once flowed on Mars raised the remote possibility of finding life there. In the distant past, surface water could have evaporated entirely or disappeared beneath the surface of the soil. Martian life may have followed the water underground and adapted itself to the changing conditions of the planet. If that was the case, then scientists could detect microbes and their biochemical by-products in the soil of Mars.

FIG. 6.3. Carl Sagan and Paul Fox superimposed the latest map of Martian canals (shown here by bold white lines) over the same area photographed by Mariner 9. They found virtually no relationship between maps and photographs of the Martian surface. (Reprinted from *Icarus*, vol. 25, Carl Sagan and Paul Fox, "The Canals of Mars: An Assessment after Mariner 9," p. 606, copyright 1975, with permission from Elsevier.)

The results of the Mariner 9 mission led scientists to lower the level of possible Martian life from lichens to microbes. Nevertheless, the possibility of finding microbes was reason enough to send costly unmanned spacecraft to land on Mars and test for their presence.

After America defeated Russia in the race to the Moon in 1969, NASA was ready for other dramatic missions using its space technology. Searching for microbes on Mars was not as newsworthy as having humans walk on the Moon, but it was the next best thing to do in space. The lunar landing of Apollo 11 captured the world's attention. Later Apollo missions created far less popular interest, especially after the examination of lunar rock samples showed no signs of life. Hence, there was a renewed effort to find life in the solar system.

Viking was the name of the mission designated by NASA to search for Martian life. The United States government spent $930 million on two Viking spacecraft, exclusive of launch costs. The design and construction of instrument packages capable of testing for life and for organic chemical compounds on Mars cost another $100 million.

It is proof of the power of extraterrestrial life on the human imagination that America was willing to spend over one billion dollars on the Viking project. The search for Martian life inspired NASA scientists and engineers to embark on the most complex and expensive scientific space mission in history. For the first time, the federal government invested a substantial amount of public money in what many scientists had long regarded a fantasy: hunting for life on Mars.

Although the Viking science teams planned to study the meteorological, seismological, chemical, and physical state of Mars, the biological experiments were the focus of their work. They were the most expensive of the planned scientific experiments, and they were the ones that drew public attention to the mission.

Viking spacecraft landed successfully on two Martian sites in July and September 1976. NASA personnel radioed instructions to the biological laboratory instruments. A mechanical arm reached out from the lander to scoop up soil and deposit it in the processing units for testing. Initially, the Viking instruments indicated that the surface of Mars was chemically or biochemically active but did not yield irrefutable proof of the existence of Martian life or organic chemical compounds. NASA scientists differed about the interpretation of the results. The experimenters were unable to reach a consensus on the subject and the debate dragged on.

Did the Viking biological and biochemical tests prove beyond any doubt that Martian life did not exist? Norman H. Horowitz, a member of the Viking biology team, argued that it was "impossible to prove that any of the reactions detected by Viking instruments were not biological in origin."[6] It was equally impossible, he added sardonically, to prove that the rocks photographed at the landing sites were not living organisms that just happened to look like rocks. Horowitz concluded that Mars was a sterile planet devoid of life and organic matter.

Five years after the failure of Viking instruments to find conclusive evidence of Martian life, Viking team leader Gerald Soffen reminisced about what they might have discovered. The orbiter camera "could have seen cities or the lights of civilization." The infrared mapper located "unusual heat sources," the water-vapor sensor found "watering holes or moisture," the entry mass spectrometer identified gases "outside the limits of chemical equilibrium," and the seismometers "detected a nearby elephant."[7] These were the nostalgic thoughts of the man who guided thirteen Viking teams responsible for the project's main scientific investigations on Mars.

Scientists committed to the idea of Martian life were not ready to accept the bleak picture of Mars that emerged from the Viking Project. The two spacecraft, critics complained, landed far from the polar caps. Perhaps life existed in frozen water near the caps? Carl Sagan and geneticist Joshua Lederberg suggested that life might thrive in warm, wet microenvironments scattered across the Martian landscape like oases in a desert. After all, the Viking landers had explored a very small portion of the planet. Still others claimed that Martian microbes had retreated underground, making it necessary to drill into the soil to uncover them.

Elusive microorganisms living within the interior of Mars are the final remnants of Lowell's grand vision of Martian life. Those remnants, however, are still capable of rousing the American public and its space agency to action. On several occasions, the prospect of finding life on Mars revived lagging public interest and gained Congressional financial support for new space efforts.

In the summer of 1996, NASA scientists claimed they had discovered fossils of microorganisms, and evidence of organic molecules, in a piece of rock ejected from Mars fifteen million years ago. The rock fragment eventually fell to Earth in Antarctica as a meteorite. Fossilized microbes in the rock supposedly dated to over three billion years ago. At that time, Mars had flowing water.

Experts questioned NASA's findings from the outset. Biologists were notably skeptical about the claims. They suspected that the so-called microfossils were inorganic chemical precipitates or the result of terrestrial contamination. Well-funded, internationally renowned laboratories took up the challenge of the Martian fossils and soon dismissed major elements of the claim that microbial life had been found on Mars. Nevertheless, this event was a great boom for the search for life on Mars and elsewhere in the Universe.

The response to the purported Martian fossils is as interesting as the discovery itself. The public quickly accepted NASA's claims. Sixty percent of Americans surveyed agreed that NASA scientists had found proof of primitive life on Mars.

In Washington, D.C., Democratic and Republican politicians immediately issued a joint call for more government spending on space science. President Bill Clinton announced a bipartisan White House meeting to discuss the U.S. space program. At this meeting, he declared his willingness to back the space agency's search for evidence of life on Mars. Clinton's promises came shortly before Congress held hearings on appropriations for future NASA projects.

The history of the changing conception of Martian life records a long retreat with supporters fighting holding actions at every point. Initially a superior race of Martians cultivated irrigated crops on the planet. Next the canal builders and their vegetation were replaced by lichens and microbes. When the Mariner missions ruled out lichens and Viking on-site biological and chemical tests failed to disclose microbes or organic molecules, there was a shift to the fossils of microbes.

Despite these setbacks, interest in Martian life remains high. Reports from the Mars Odyssey spacecraft in the spring of 2002 show that significant quantities of frozen water exist at the planet's north pole, along with frozen carbon dioxide. The presence of water on the planet has raised hopes once more that small-scale life inhabits Mars.

In January 2004, the European Mars Express detected ice at Mars's southern pole. Shortly thereafter, NASA orbited spacecraft around Mars and landed two rovers on the planet. In the midst of this renewed interest in Mars, President George W. Bush announced two long-term goals for the space agency: build a base on the Moon and send astronauts to Mars. Subsequent studies of Mars confirmed that water once existed there and that the Martian atmosphere contains detectable amounts of methane, the simplest hydrocarbon molecule.

Lowell's Successor: Carl Sagan

The space age ended Lowell's influence on the image of Mars as an inhabited planet. Simultaneously, it gave birth to Lowell's successor as the foremost champion of life on other planets: Carl Sagan (1934–1996). Sagan was a planetary astronomer and public figure who often criticized Lowell, yet his career bears a remarkable resemblance to Lowell's. The two were the best-known advocates of extraterrestrial life during their lifetimes. They deliberately and tirelessly courted public approval of their scientific work. Finally, they suffered criticism from fellow scientists who believed that their theories and public relations tactics sometimes went beyond the limits of good science.

Substantial differences existed between the two men. Sagan was the more gifted scientist of the pair, and he was less dogmatic about his theories than Lowell. However, Sagan's scientific side was at war with an unrestrained speculative streak that ran through his work. In addition, Sagan practiced astronomy in an era when the financial support of the federal government, and not the generosity of wealthy donors, was the primary source of research funds for the astronomical sciences. Despite these important differences, similarities persisted between the two astronomers. A colleague of Sagan's once remarked, "If Lowell hadn't existed, Sagan might have invented him."[8]

Sagan studied astronomy with Gerard P. Kuiper, the pioneer planetary astronomer who thought that simple lichens or moss might live on Mars. Sagan's early interest in the origins of terrestrial and extraterrestrial life led him to pursue additional study in genetics and biochemistry. He spent the summer of 1952 at the University of Indiana working in the laboratory of the Nobel Prize winning biologist H. J. Muller. Unlike many proponents of extraterrestrial life, Sagan came to the subject with a thorough grounding in the physical and biological sciences. His first published scientific paper attempted to explain how living organisms arose from organic molecules.

Sagan made two important contributions to planetary science early in his career. He and his colleague James Pollack proposed that the color changes that swept periodically over Mars were due to dust storms and not to the spread of vegetation. Sagan was also among the first to claim that the high surface temperature of Venus was due to a greenhouse effect that retained solar heat and to the presence of water vapor in the Venusian atmosphere.

In 1966 Sagan publicly declared his interest in extraterrestrial life when he collaborated with the Soviet astrophysicist Iosof S. Shklovskii to produce *Intelligent Life in the Universe.* This book is an enlarged English-language version of an earlier work written by Shklovskii and published in the Soviet Union.

The volume is remarkable for two reasons. Sagan and Shklovskii were the first major modern astronomers to endorse the search for advanced extraterrestrial life. And, the collaborative authorship of an American and Soviet scientist was unusual in the Cold War era.

Intelligent Life in the Universe first summarized the physical structure of the universe. It then covered the origins of life and the possible existence of intelligent extraterrestrial life. A photograph in the volume depicts a starcloud consisting of over one million stars. The caption under the illustration assures the reader that one of these stars nurtures a technological civilization superior to ours.

Sagan was largely responsible for the chapter dealing with life on Mars. He presented a fair summary of the relevant evidence for Martian life and dismissed Martian canals as psychological and visual illusions. Sagan chastised Lowell for hampering the development of astronomy by needlessly prolonging the canal debate. However, as Sagan neared the end of the chapter, he depicted a Mars that was similar to the one made famous by Lowell.

Sagan proposed that the space agency deliver a computer-controlled biological laboratory to Mars to gather information on the physical and biological state of the planet. It would be advisable, Sagan noted, to incorporate a television camera into the automatic laboratory to scan the Martian landscape at regular intervals. He admitted that the scanned images would likely be predictable—rocks, lava flows, sand dunes—and then he added: "An occasional scraggly plant would not be unexpected. But there are other possibilities—fossils, footprints, minarets."[9]

How could Sagan put footprints and minarets on a planet that scientists thought *may* contain lichens and microbes? Was he writing about the Mars portrayed in modern astronomy or about Barsoom in the science fiction novels of Edgar Rice Burroughs? Throughout his career, Sagan made spectacular claims of this sort. They fueled the criticisms of his detractors, pleased a public eager to hear about exotic alien life forms, and revealed Sagan's strong belief in advanced extraterrestrial civilizations.

Sagan did not stop with the minarets of Mars. His co-author inspired him to even wilder speculations about Martian civilization. Shklovskii earlier had drawn attention to anomalies in the orbits of Phobos and Deimos, the twin moons of Mars. At that time, astronomers could not explain these orbital discrepancies using available physical evidence. This led Shklovskii to leap to the extraordinary conclusion that the Martian moons were not natural.

He alleged that the eccentricities of their orbits indicated that the moons were actually gigantic hollow satellites constructed by Martians and placed in orbit around their planet. The highly civilized Martians filled their spacious satellites with libraries and museums recording the extraordinary history and accomplishments of their doomed culture. Shklovskii concluded that an advanced Martian civilization would have placed artificial satellites in orbit when it first ventured into space.

Shklovskii presented this radical hypothesis in 1959, two years after the Soviet Union orbited the first artificial satellite (Sputnik) around the Earth. He was probably inspired by the spectacular space successes of his country when he claimed that Phobos and Deimos were artificial satellites like Sputnik. The Russian version of Shklovskii's book appeared on the fifth anniversary of the launching of Sputnik.

Sagan embraced Shklovskii's idea and went on to calculate that a satellite the size of a Martian moon would weigh tens of millions, or billions, of tons. He decided it might be easier to hollow out a small asteroid than to construct an orbiting satellite with material brought from the surface of the planet. Sagan did not explain how Martians excavated a rocky asteroid several miles in diameter. He left his readers to suppose that converting an asteroid to an artificial satellite is one of those things that superior extraterrestrials do more easily than humans.

The possibility of artificial Martian moons led Sagan to speculate about their makers. Martian civilization, he wrote, must have been "mighty indeed"[10] if hundreds of millions of years ago it was able to place a gigantic satellite in orbit about its planet. Where is that civilization now, he asked. We cannot answer that question, Sagan acknowledged, until the first exploration teams from Earth visit Mars.

Why did Sagan write about Martian minarets and artificial satellites circling Mars when he knew that most scientists believed only simple life forms, at best, could live on the planet? Sagan knew better than to expect to glimpse a minaret on Mars. At the same time, he realized that popular support for costly missions to Mars depended upon raising public expectations of finding life on the planet.

Sagan was not seeking personal gain when he discussed intelligent life on Mars. He passionately believed that the possibility of the existence of extraterrestrial life was the most important question confronting humanity. It was so important that it might serve as an alternative to war and bring eternal peace to Earth. "More effort up there," wrote Sagan, "less chance of fighting down here."[11] Therefore, the U.S. government, with public support, must continue funding space ventures.

When Percival Lowell catered to popular audiences by writing about Martian canals and their builders, astronomers accused him of taking "the popular side of the most popular scientific question" of the time. "The world at large is anxious for the discovery of intelligent life on Mars," his critics complained, "and every advocate gets an instant and large audience."[12]

More than a half century later, Sagan followed the path of advocacy opened by Lowell. Life on Mars was a very popular topic in space age America. An enterprising scientist could gain public recognition and financial support for research by suggesting that extraterrestrial life might exist somewhere in space. Sagan obtained research grants, both military and civilian, and served on many space boards and commissions.

There is another explanation for Sagan's extravagant claims for advanced Martian life. Sagan freely admitted the lasting impact Burroughs's Martian novels had on his imagination. Burroughs was the first person to locate minarets on Mars. In his novel *A Fighting Man of Mars,* Burroughs wrote that "towers, domes and minarets"[13] are typical architectural features of Barsoomian cities. When discussing Mars, Sagan repeatedly quoted Burroughs's phrase, "beneath the hurtling moons of Barsoom." In 1971 the Mariner 9 spacecraft televised the first close-up pictures of Phobos. Sagan and a NASA technician were the first to view the surface of the Martian moon. In honor of this occasion, the state of California issued Sagan an automobile license plate marked PHOBOS. He preferred one that read BARSOOM, but the state of California limited vanity plates to six letters.

The first detailed images of Phobos did not shake Sagan's confidence in Shklovskii's hypothesis. In his 1973 book *The Cosmic Connection*, Sagan lists the hypothesis as one of three viable scientific explanations for the origins of the Martian moons. The other two hypotheses are that the moons were asteroids captured by Mars's gravitation or debris left over from the formation of the planet.

It was relatively easy to transform Barsoom's hurtling moons into a pair of artificial satellites. In his analysis of Shklovskii's satellites, Sagan noted that Burroughs's account of the motion of Barsoom's moons was wrong. The moons do not hurtle. They barely creep across the Martian sky. Sagan's frequent references to Burroughs do not prove that he was unable to separate science fiction from science fact. Instead, they show the depth of his intellectual and emotional commitment to the idea of extraterrestrial civilizations.

It is worthwhile to examine the claim that the moons of Mars are artificial satellites. Phobos and Deimos are very small irregularly shaped bodies, roughly 13.2 and 7.2 miles in diameter, that move in orbits close to Mars. Our Moon, by comparison, has a diameter of over 2,000 miles and is 237,000 miles distant from the Earth.

At the time Sagan and Shklovskii were writing their book, telescopic observers saw Phobos and Deimos as two specks of light traveling near the planet. Information about these tiny satellites, gathered at a distance of 40 million miles, disclosed that their orbits were decaying. Astronomers predicted that the Martian moons would crash onto the planet sometime in the future.

Shklovskii used these predictions to prove that the moons were artificial satellites. After all, Sputnik was in a decaying orbit and it too would plunge toward the planet it circled. Later observations of the Martian moons have not supported Shklovskii's hypothesis. Astronomers concluded that the moons are solid bodies, probably asteroids captured by the gravitational field of Mars. Eventually, Sagan accepted their findings. He compared the moons to "old, battered-up rocks."[14]

In the book he wrote with Shklovskii, Sagan rejected the "bizarre suggestion" of Soviet writer F. Zigel, who claimed that Martians launched their artificial satellites in 1877. This was close to the time when Asaph Hall first observed the moons of Mars. Zigel's scenario explained why Hall discovered the moons in 1877, and it implied that a thriving space program might currently exist on Mars. Dismissing Zigel's hypothesis, Sagan concluded that Martian moons "are much more likely mute testaments to an ancient Martian civilization than signs of a thriving contemporary society."[15] Sagan's suggestion of an ancient Martian civilization is hardly less bizarre than Zigel's claim that Martians were busy launching satellites in 1877.

The Rational Speculator

Carl Sagan spent his life attacking frauds and debunking fabulous flying saucer stories. In one of his last books, *The Demon-Haunted World* (1996), he exposed the many hoaxes, myths, and superstitions that plague the world today. He made an eloquent plea for a return to rationality in an age when New Age prophets, religious fundamentalists, and promoters of various pseudosciences threaten the growth of science.

Sagan saw science as our best hope, a candle of rationality shining in darkness and chaos. However, he was willing to excite the public with tales of ancient Martian civilizations and million-ton satellites if such stories encouraged exploration for extraterrestrial life. Sagan's comments about ancient Martian civilizations are as suspect as many of the fraudulent claims he exposed in *The Demon-Haunted World*. His friend Bruce Murray claimed that although Sagan understood that Mars was probably lifeless, he had relapses when he would discuss Martian life. "If anything," continued Murray, "that was his UFOs."[16]

Sagan's ambiguous response to the notorious "Face on Mars" reveals the contending forces at work in his mind. In 1976 Viking 1 orbiter, traveling over the Cydonia region of Mars, captured a low-resolution image of what appeared to be a monumental human face. Initially, NASA scientists dismissed the image as an optical illusion, a chance conjunction of light and shadow on the planet's surface. Nevertheless, NASA issued a photographic image with the caption "Face on Mars?" By 1984 the photograph was featured in the tabloid newspaper *Weekly World News*, and in 1987 Richard Hoagland, a former NASA consultant, published a book, *The Monuments of Mars*. According to Hoagland's book, the Face was surrounded by remains of a city, fortress, and pyramids. Sagan responded with an article in *Parade*, a magazine issued as a supplement to Sunday newspapers. He dismissed the Face and its associated structures as fantasy, not science.

In early 1998, NASA prepared targets on Mars for imaging by the Mars Global Surveyor (MGS). Supporters of Hoagland's grand theory of the Face appealed to NASA. They asked the agency to make the Cydonia region and its Face one of the imaging sites. Michael Malin, the NASA scientist in charge of space cameras on MGS, resisted the suggestion. He said it was a waste of money, and maneuvering the spacecraft to get another image of the supposed Face might endanger the entire mission.

Malin was overruled by NASA officials who bowed to public pressure for a picture that was worthless from a scientific standpoint. As Malin battled with NASA officials, he was surprised to learn that Sagan had changed his mind about the affair. *In The Demon-Haunted World*, Sagan wrote that the Face on Mars was probably natural, not artificial, but it deserved further study because the hypothesis posed by its supporters belonged in the scientific arena. Once more Sagan allied himself with popular opinion and opposed the consensus of scientists in the matter. As for the Face, it did not appear on the high-resolution images recorded by the Mars Global Surveyor.

Sagan alone cannot be blamed for his continuing interest in the possibility of extraterrestrial intelligence despite the skepticism and opposition of many scientists. Sagan's colleagues welcomed his earlier work, *Intelligent Life in the Universe*, as a valuable contribution to an emerging scientific field. The volume received endorsements from biologist H. J. Muller and astronomers Frank Drake and Fred Whipple. Several other scientists called attention to this work on the occasion of Sagan's sixtieth birthday celebration in 1994. They praised it as a landmark book that prepared a new generation of seekers of extraterrestrial intelligence. Historian Steven J. Dick rightly called the volume "the bible of scientific thought on extraterrestrial life."[17]

Sagan studied the long history of scientific speculation about extraterrestrial life and civilization. His reading in the early scientific literature on the subject led him to Huygens's *Kosmotheoros* of 1698. Sagan admired the Dutch astronomer's description of a universe teeming with life: "So many Suns, so many Earths, and every one of them stock'd with so many Herbs, Trees and Animals."[18] Despite his admiration for Huygens, Sagan criticized him for naively transplanting the physical environment and inhabitants of Earth in the seventeenth century to the planets. He tempered his criticism with the observation that "Huygens was, of course, a citizen of his time." Then he added, "Who of us is not?"[19]

Sagan recognized that Huygens was caught in the web of his times. He did not see that he was the victim of similar circumstances when he theorized about advanced life on Mars. His guide to *Intelligent Life in the Universe* reflects the ideas and enthusiasms of the 1960s. These were the early years of the space program when journalists, politicians, and NASA publicists freely used the maritime discovery analogy. They believed that just as Christopher Columbus's ships carried him to the New World, so would spacecraft reveal new worlds in space. The Soviet-American race to the Moon, driven by Cold War competition for supremacy in space, recalled European rivalry over the New World in the sixteenth century. Voyages to the Moon were the first steps in exploratory programs intended to carry humans to Mars and beyond. Sagan, in keeping with the spirit of the times, wrote about an ancient Martian civilization awaiting discovery by intrepid space navigators.

Sagan knew that the principle of mediocrity had its limits. He understood that its claim that the rest of the universe was similar to the portion we study was more useful to astronomers and physicists than to biologists and seekers of intelligent extraterrestrial life. Therefore, Sagan highlighted the fallacies generated by biological chauvinism, the erroneous belief that life elsewhere in the universe must be essentially the same as life on Earth.

Sagan attacked carbon- and oxygen-chauvinism, the belief that life cannot exist without these elements, and noted the random and contingent nature of evolution. Genetic mutations and the recombination of genes introduce randomness into the evolutionary process. The process is contingent because chance plays an important role in it. Chance in the form of rare and catastrophic disasters has led to the extinction of some forms of life and the expansion of others.

Sagan maintained that the forces of evolution operating in distant parts of the universe would not generate creatures exactly like us. As a technical consultant for Stanley Kubrick's film *2001: A Space Odyssey*, Sagan advised against the portrayal of extraterrestrials in any form. He argued "that nothing like us is ever likely to evolve again anywhere else in the universe."[20]

Decades before Sagan coined the term "biological chauvinism," Percival Lowell exposed the fallacy of thinking that advanced extraterrestrial life must assume a human form. Lowell made his point by drawing upon the biological sciences. He argued that lungs are not absolutely necessary to life and that nothing prevented a gill-breathing creature "from being a most superior person."[21] A fish, he continued, might imagine that it is impossible to live without water. Likewise, many believe that advanced life forms cannot exist on Mars because the Martian atmosphere is thinner than the Earth's. Lowell concluded that anyone who accepts this type of argument is not thinking like a philosopher but like a fish. Lowell's Martians may have been engineers, inventors, administrators, and workers, but they did not necessarily share the same physical features as humans beings.

Sagan exposed the parochialism of biological chauvinism at the same time that he adhered to other forms of chauvinistic thought. He was willing to consider radically different chemical bases for life, but assumed that crucial features of human culture are duplicated in the universe. Sagan's extraterrestrial beings may be free of the need for carbon or oxygen, but, like Lowell's Martians, they establish civilizations and cultivate technologies copied closely after those found on Earth. Lowell and Sagan represent different eras and different views of science, yet they shared visions of a great Martian civilization.

CHAPTER SEVEN

Carl Sagan

Mars and Beyond

> We have launched four ships to the stars, Pioneers 10 and 11 and Voyagers 1 and 2. They are backward and primitive craft, moving, compared to the immense interstellar distances, with the slowness of a race in a dream. But in the future we will do better. Our ships will travel faster. There will be designated interstellar objectives, and sooner or later our spacecraft will have human crews.
>
> —Carl Sagan, *Cosmos*, 1980

Exobiology

In his sophomore year at the University of Chicago, Carl Sagan studied chemistry with Nobel laureate Harold Urey. At about the same time, Urey's student Stanley Miller began his experiments on the origins of life. Miller mixed together hydrogen, methane, ammonia, and water vapor, the gases believed to compose the Earth's atmosphere at the emergence of life. He then simulated the effects of lightning by passing electrical discharges through this gaseous mixture. Miller, of course, did not create life in his laboratory. He showed that chemical reactions similar to those that may have occurred on the primitive Earth could yield complex organic molecules.

Miller's landmark work, carried out in 1953, meant that scientists could study the origins of life in a laboratory setting. Urey and Miller were interested in terrestrial life, but others soon explored the cosmic implications of their work. Life might appear in other regions of the universe where Earthlike conditions prevailed. It was no longer an exclusively terrestrial phenomenon.

Carl Sagan started graduate work at the University of Chicago in the midst of these new developments on the origins of terrestrial and extraterrestrial life. Sagan's mentor at Chicago was the astronomer Gerard Kuiper. While completing his doctoral studies with Kuiper, Sagan met the Nobel laureate geneticist Joshua Lederberg (1925–). The two men were brought together by their common interest in the origins of terrestrial life and the possibility of extraterrestrial life.

Lederberg was a pioneer in the scientific study of extraterrestrial life. In 1960 he named the new science *exobiology* and used his considerable scientific reputation to enhance its credibility. Lederberg hoped that exobiology would give America's space program a new focus. Instead of concentrating upon missiles and manned flight, the program would turn to scientific topics.

From its beginnings, exobiology was a highly speculative and controversial field of study. Despite its uncertain status among scientists, exobiology found a home within the space sciences. NASA created an Office of Life Sciences in 1960, sponsored conferences on extraterrestrial life, and funded exobiological research.

Upon Lederberg's recommendation, Carl Sagan was asked to serve on a number of government panels and commissions that advised NASA on matters relating to space exploration and biology. Sagan worked with NASA during the directorship of James C. Fletcher. Fletcher was a very able administrator, a tireless advocate for space exploration, and a supporter of the search for extraterrestrial life. He was also a devout Mormon whose religion had long taught that inhabited worlds existed outside our solar system. For these varied reasons, Fletcher brought the Viking Mars missions to a successful completion and encouraged NASA-sponsored efforts to communicate with intelligent beings outside the solar system.

Joshua Lederberg's and Carl Sagan's strong belief in the existence of extraterrestrial life found favor in some NASA circles, but it was disputed by officials and scientists preparing for the Apollo flights to the Moon from 1969 to 1972. The central issue of the dispute, microbial contamination, pitted exobiologists against geologists and engineers at the space agency. The exobiologists warned that terrestrial microbes carried to the Moon by Apollo spacecraft and astronauts could endanger lunar life forms. Likewise, lunar microbes accidentally brought back from the Moon might infect inhabitants of the Earth.

Sagan urged the sterilization of all spacecraft traveling to the Moon. NASA officials ruled out sterilization of the Apollo lunar landers. It would have been very difficult, if not impossible, to place a sterile lander on the Moon. In order to protect humans from infection by lunar microbes, NASA agreed to quarantine returning Apollo astronauts and their cargo of lunar rocks.

Appropriate tests made on the first astronaut team to visit the Moon showed no evidence of lunar life. NASA dropped the quarantine of returning astronauts after several more lunar missions. By 1972, when Apollo 16 flew to the Moon, virtually all scientists agreed that the Moon was lifeless. Carl Sagan was one of the few to dispute this conclusion. He maintained that microorganisms might live deep beneath the lunar surface.

NASA's plans to land instruments on Mars as part of Viking missions 1 and 2 reopened the issue of terrestrial contamination in the mid-1970s. Lederberg joined Sagan in calling for the sterilization of the Viking landers and modification of the descent rocket engines so that their nozzle blasts would not destroy any Martian life in their paths. The exobiologists won that round of arguments. NASA heat-sterilized the two Viking landers and fitted shields onto the sterilized spacecraft to protect the landers from contamination until they reached Mars.

Norman Horowitz, a biochemist at California Institute of Technology, did early work on the chemical origin of life. Although he later served as one of the chief biologists on the Viking Project, Horowitz was skeptical about the existence of Martian microbial life. He wrote that spacecraft sterilization was "a monument to a Mars that never existed."[1] Horowitz was referring to Mars as conceived by Lowell and popularized by science fiction writers. The dangers of microbial contamination figured prominently in fictional encounters between Martians and humans. Common terrestrial microorganisms killed the invading Martians of H. G. Wells's *War of the Worlds* (1897) and the innocent Martian natives of Ray Bradbury's *Martian Chronicles* (1950).

In the early 1970s, Sagan became a member of NASA's imaging teams for Mariner 9 and the two Viking missions. The imaging team interpreted the pictures of the Martian surface recorded by the spacecraft's electronic cameras. Sagan was the sole astronomer, and the only scientist with an extensive background in biology, on the team. Furthermore, he was a staunch believer in the existence of advanced Martian life at a time when most scientists considered it doubtful that microbes existed on the planet. One of his NASA colleagues explained Sagan's position in these words: "Sagan struggles to create situations where life might exist. It's a compulsion."[2]

Apart from the part he played in the campaign for spacecraft sterilization, Carl Sagan exerted considerable influence on the planning stages of the Viking missions. He helped in the placement and operation of the video cameras designed to obtain pictures at the Martian landing sites. Sagan believed that these cameras were more useful for detecting life on the planet than the expensive automated bio-testing equipment carried to Mars.

While Viking biologists concentrated on the search for microscopic life and traces of organic compounds in Martian soil, Sagan prepared to take pictures of the *macrobes* of Mars. Macrobes, a term coined by Sagan, are forms of life visible to the naked eye or the lens of a camera. They can be shrubs, trees, or animals. With macrobes as his target, Sagan argued that the cameras should be mobile so that NASA controllers could move them to an organism situated at a distance. Sagan did not get his wish for tractor-mounted cameras, nor was he successful in his request for a light to illuminate any creature that might stroll by the landers in the dark. But he was able to have the landers' cameras modified to capture images of moving objects.

Sagan argued that Martian life was macrobial by citing the same reasons that Viking biologists used to argue it was microbial. The biologists reasoned that microbes alone could live on a planet with a very low temperature, scarce water, and low atmospheric pressure. Sagan countered that organisms facing harsh conditions on Mars might adapt by growing large bodies to conserve heat and water more efficiently. As proof he cited polar bears, which thrive in the extreme environment of the Earth's Arctic region. There was fictional, as well as scientific, backing for polar bearlike creatures on Mars. Edgar Rice Burroughs populated the icy arctic wastes of Barsoom with the ferocious *apt*, a huge white-furred animal with six limbs.

Sagan used a full-page photograph of a polar bear to illustrate his discussion of Martian macrobes in his book *Other Worlds* (1975). The image of the polar bear appears two pages after an illustration of a *banth*, a large Barsoomian lion. The caption under the picture of the banth reads, "COME TO BARSOOM." Sagan extended his invitation to visit Barsoom shortly before the Viking missions left for Mars.

Sagan and Joshua Lederberg informed the readers of *Time* magazine what forms large animals might assume on Mars. There were Crystophages (ice eaters), who got water from the planet's permafrost, and Petrophages, who fed on Martian rocks. Fanciful illustrations of these Martian beasts accompanied the article. The ice eater resembled an organic version of a flying saucer, and the rock eater sprawled across the Martian landscape in the shape of a giant octopus.

The two Viking landers touched down on Mars in July and September 1976. The landers' bio-detection laboratories initially recorded strong positive responses. Viking biologists later determined that the responses were chemical, not biological. Inorganic compounds in the Martian soil had reacted chemically with test reagents sent from the Earth. Additional tests failed to disclose either living organisms or organic molecules on the planet. The Viking scientists finally concluded that the environment of Mars was hostile to organic matter.

Carl Sagan was skeptical of the interpretation of evidence gathered by the Viking landers. In 1977 he obtained a small grant from NASA so that his students could examine thousands of images gathered by the Viking orbiters as they circled Mars at a low altitude. Sagan and his students studied the Viking orbiter images looking for proof of an advanced Martian civilization. He believed that NASA's concentration upon Martian microorganisms might have led the space agency to overlook large Martian constructions. Sagan's group found no traces of a present or past technological civilization on Mars.

Sagan's scrutiny of the Viking pictures was his last effort to uncover Barsoom on the new Mars revealed by space technology. The failure to discover Martian life at any level was a setback for Sagan and the new science of exobiology. In the fall of 1979, a group of interested scientists met to discuss the implications of the failure to discover signs of life beyond the Earth. They attended a conference appropriately named "Where are They?" Carl Sagan was not present at this meeting.

The Showman

Sagan's single-minded pursuit of extraterrestrial life and his readiness to accept any possibility that complex organisms inhabited Mars alienated many who worked with him at NASA. His loose theorizing about the formation of the Martian landscape troubled NASA geologists. For example, Sagan proposed that a mere twelve thousand years ago Mars was a wet, warm planet. He envisioned a world with "balmy temperatures, soft nights, and the trickle of liquid water down innumerable streams and rivulets."[3] Geologists responded that if there was ever such a Mars, it existed billions, not thousands, of years ago.

Sagan often relied upon the double negative when he wrote about extraterrestrials. It is a rhetorical trick long favored by some defenders of life on other worlds. Schiaparelli made effective use of the double negative in the 1890s. It permitted him to make positive sounding statements about intelligent Martian life without committing himself to it. Asked if the *canali* on Mars were artificial, Schiaparelli replied: "I am very careful not to combat this supposition, which includes nothing impossible."[4]

In a similar vein, Sagan responded to questions about large animals on Mars by saying, "there is no reason to exclude from Mars organisms ranging in size from ants to polar bears."[5] When asked about his interpretation of the close-up images sent by Mariner 9, Sagan asserted there was "nothing in these observations that excludes biology."[6] Sagan used double negatives in his technical papers and in his popular writings. The usage is grammatically correct, but the statements are evasive and confusing. NASA biologist Harold Klein said that Sagan favored statements of the form: "We have seen nothing that can rule this out." Klein added, "He's very clever—he doesn't promise anything."[7]

Sagan's evasive statements, outrageous theories, and haughty behavior exasperated his NASA colleagues. However, many of them came to value him as a useful gadfly who forced them to expand their ideas of life beyond the Earth. Eventually, Sagan's critics conceded that he played an important role in maintaining public support for space science during the lean years following the end of the Apollo Moon voyages. Sagan discussed the space program within a broad philosophical context. He talked about the religious, social, and political implications of space at a level the average American understood.

Sagan's public campaign for the sterilization of the Viking landers, and his provocative remarks about Martian life, helped to promote popular interest in the expensive Viking missions. Sagan better understood the public's response to America's space efforts than many of his NASA colleagues. It was the search for extraterrestrial life, not the overall advancement of scientific research, that roused the populace to support NASA projects. From that perspective, it was more productive to speculate about Martian minarets and polar bears, or a lost Martian civilization, than to investigate the geological history of the planet.

The career of Bruce Murray serves as a case study of a colleague's changing response to Sagan's dual role as scientist and public promoter of space exploration. Murray, a member of the faculty of the California Institute of Technology, specialized in the geology of Mars. In 1971 Murray was worried that the prevailing Lowellian conception of Mars would distort the interpretation of the first close-up pictures of the planet taken by Mariner 9. His misgivings persisted as NASA made preparations for the Viking missions to Mars. Murray believed that it was wrong to search for clues of Martian life before the physical environment of the planet was thoroughly studied and understood. He candidly wrote in his autobiography, "Postulating the existence of Martian life publicly as a means of developing support for Viking made me uncomfortable." "I was a critic, not a supporter, of Viking,"[8] he admitted. Murray was clearly at odds with Sagan's approach to planetary exploration.

In April 1976, three months before the first Viking landing on Mars, Murray became the director of the Jet Propulsion Laboratory. JPL is a research facility operated by Cal Tech and funded by government contracts. No longer an academic scientist, but a NASA administrator, Murray soon found it expedient to use "Viking's visibility at JPL to sell new planetary missions."[9] Three years later Murray joined Sagan to create The Planetary Society, an organization that promotes planetary exploration and the search for extraterrestrial life. Murray's experience was not unique. More and more NASA scientists, engineers, and administrators came to realize that it was the promise of life on other worlds that excited public interest in space missions.

By the mid-1970s, Carl Sagan was not only the most prominent defender of extraterrestrial life, but he was also the best known scientist in America. Parallels between Sagan and Lowell immediately come to mind. Sagan, however, had media technologies available to him that were unknown to Lowell. Sagan made use of cheap paperback books, movies, and television to advance his belief in life beyond the Earth, gain support for space science, and promote himself as a public figure.

Sagan's popular books *The Cosmic Connection* (1973) and *Other Worlds* (1975) sold millions of copies to readers enthralled by the exploration of space. However, he gained his first great triumph as a popularizer of science doing guest appearances on Johnny Carson's *Tonight Show*. The comedian was host of the most popular late-night television program in the 1970s. Even though Sagan's scientific colleagues complained about his unabashed showmanship, they admitted that it was showmanship in the service of science and the space program.

Message to the Stars

Sagan began his scientific career as a planetary astronomer. His continuing interest in advanced extraterrestrial life led him into work that enhanced his popularity and soon transcended the boundaries of our solar system.

In 1970 Sagan persuaded NASA to sponsor a project to send the first message to the inhabitants of interstellar space. This effort, which was part of the Pioneer 10 space mission, displayed Sagan's superb skills as a publicist for space science. There was little chance that the messages would reach intelligent aliens, but their formulation and dispatch stirred interest around the world.

NASA scheduled Pioneer 10 for launching on March 1972. The spacecraft was scheduled to travel through the asteroid belt lying between Mars and Jupiter and on to Jupiter and its system of moons. After gathering data near Jupiter, the spacecraft would follow a trajectory that took it beyond the solar system. Pioneer 10 was the first human artifact designed to cross the outer limits of our solar system and enter interstellar space.

The historic significance of the Pioneer mission was immediately evident to science writer Eric Burgess, who suggested that Pioneer 10 carry a message from humanity to intelligent beings in outer space. He argued that because Pioneer 10 was the first artificial object to leave the solar system, it was a sign to the rest of the universe that an advanced technological civilization existed elsewhere in space.

Sagan and radio astronomer Frank Drake immediately seized upon Burgess's idea. With the help of Sagan's artist-wife Linda, they designed a small plaque for the spacecraft. This plaque carried a coded message for extraterrestrial beings who might capture the space vehicle millions or billions of years in the future.

Earlier plans to communicate with intelligent aliens by signal lights or radio contrasted with the simple, cheap, and feasible proposal submitted by Sagan and Drake. They could deliver an interstellar message by merely attaching a 6 x 9-inch wafer-thin, gold-anodized aluminum sheet to the strut of a spacecraft whose astronomical mission NASA had established earlier. It was necessary, of course, for NASA officials to evaluate this proposal. NASA director James Fletcher immediately approved the Sagan-Drake project.

Apart from its historical significance as the first terrestrial message sent to intelligent life beyond the solar system, the Pioneer 10 plaque also had implications for those who remained on Earth. NASA's acceptance of the plaque was proof of its support for the idea that advanced technological civilizations existed in space and that it was worthwhile to attempt to reach them. The message plaque encouraged the adoption of a cosmic perspective among the general public and enhanced the credibility of advocates of extraterrestrial communication.

Working under a severe time limit, the Sagan team made decisions that opened them to criticisms from several quarters. Sagan acknowledged that the hurriedly assembled message inadvertently contained material perhaps better understood by humans than extraterrestrial creatures. Nevertheless, he was confident that an advanced technical civilization could overcome the human bias in the message and decipher it easily.

The primary aim of the plaque was to inform its finders about the nature, location, and epoch of the makers of the spacecraft. Sagan accomplished this by using basic scientific information presumed available to any advanced civilization. He was absolutely certain of two things. First, the laws of physics held true wherever Pioneer 10 might travel. Second, because of their universality, science and mathematics developed on Earth would be understood by other intelligent creatures in the universe.

The Sagan team located the key to decoding Pioneer's message at the top left portion of the plaque (Fig. 7.1). The two adjacent circles joined by a straight line is a schematic representation of the hyperfine transition of neutral hydrogen. A magnetic interaction between the electron and proton of a hydrogen atom causes it to emit radio waves with a wavelength of twenty-one centimeters. This became the basic unit of distance needed for unscrambling the remainder of the

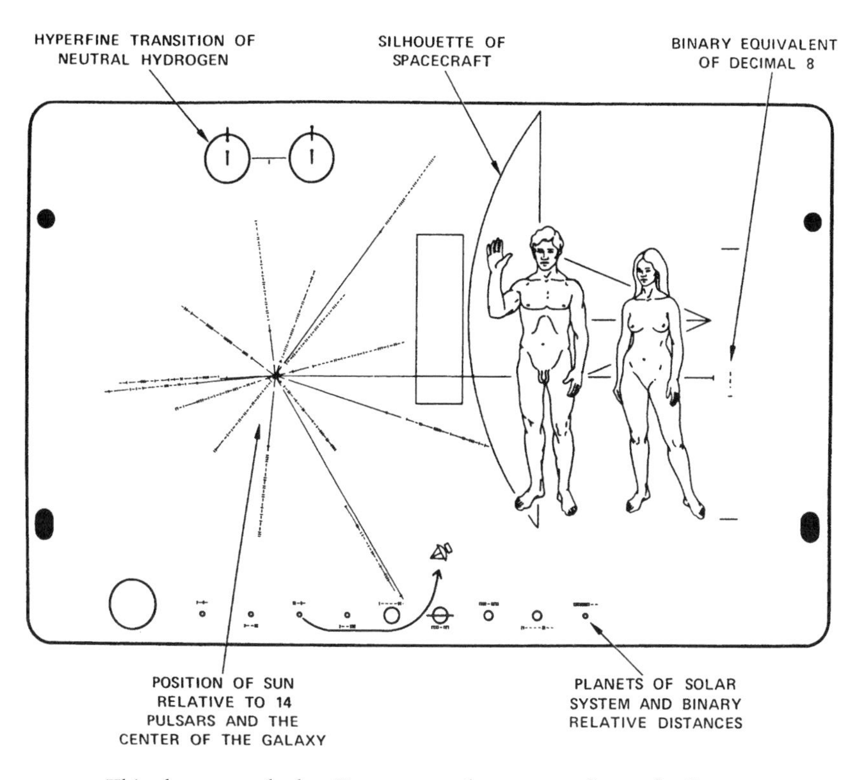

FIG. 7.1. This plaque, attached to Pioneer 10 and 11 spacecraft, was the first attempt to send a physical message beyond our solar system. (Richard. O. Fimmel, James van Allen, and Eric Burgess, *Pioneer: First to Jupiter, Saturn, and Beyond*. Prepared at Ames Research Center. Washington, D.C.: NASA, 1980.)

message. Hydrogen, Sagan notes, is the most abundant element in the Galaxy, and therefore the hyperfine transition pictured on the plaque "should be readily recognizable to the physicists of other civilizations."[10]

Sagan placed the binary equivalent of the number eight on the far right of the plaque. It lies between two tote marks indicating the height of a nude female of our species. Sagan believed that an advanced society, one capable of retrieving a passing spacecraft, could determine that eight times twenty-one centimeters is the height of the depicted female. This number also allows the finder to verify the dimensions of Pioneer's silhouette, drawn to scale behind the naked figures.

The next most important feature of the message is the radial pattern of dashed lines at left center. These lines form a map of fourteen pulsars. Pulsars, first named by Frank Drake, are spinning neutron stars emitting pulses of radio energy. The Pioneer 10 plaque locates the Sun relative to the depicted pulsars and the center of our Galaxy. On each of the lines is a string of ten-digit numbers in decimal notation representing the time interval between successive radio emissions of the pulsars at the spacecraft's launch date. Because pulsars run down at known rates, they serve as accurate galactic clocks for hundreds of millions of years. A scientifically sophisticated society that has maintained records of pulsar rates should be able, Sagan argued, to determine the precise time Pioneer 10 left Earth. The message plaque as conceived by Sagan and Drake provides sufficient information to specify a single star, the Sun, in 250 billion stars and one year, 1970, in 10 billion years.

At the bottom of the plaque, Sagan portrayed the solar system schematically. The Sun is at the extreme left, and Pioneer's trajectory is the curved line that starts at the Earth and bends along Jupiter before heading away from the solar system. Artists sketched the naked human messengers with mixed racial features. The taller, male figure has his right arm raised in a gesture of good will. His hand is open to show he carries no weapons. Sagan accented the opposable thumb on the figure's hand. This anatomical feature was believed crucial in setting humans apart from apes and permitting them to develop the tools that enabled them to create a technological civilization.

Pioneer 10 was launched from Cape Kennedy on March 3, 1972. According to Sagan's calculations, it would not enter the planetary system of a star in our Galaxy within the next ten billion years. Hence, Sagan likened the message carried by Pioneer to a note sealed in a bottle and tossed into the ocean for eventual retrieval by a passing ship. In Pioneer's case, Sagan hoped that a civilization technologically superior to ours might have the means to detect the spacecraft in interstellar space, determine its artificial origin, and then retrieve it for study. NASA further endorsed Sagan's handling of the project by attaching a similar plaque to its next interstellar craft (Pioneer 11) launched in April 1973.

Critics in America and Europe carefully scrutinized the contents and code of this historic message to the stars. Some argued that a larger and more diverse committee should have determined the contents of the communication. Others said that the coded message was so obscure that trained physicists and astronomers on Earth had difficulty reading it. The noted art historian E. H. Gombrich complained that the pictorial aspects of the message assumed that extraterrestrials shared the conventions Western viewers used to decode the picture. The depicted human figures presented problems in pictorial perspective, he argued, and an understanding of the spacecraft's trajectory depended upon a directional arrow, a symbol meaningful only to societies familiar with bows and arrows. Sagan countered that his team did its best given the short time allotted for the completion of the task.

A smaller number of critics commented on the most basic assumption of the message: that superior extraterrestrial technological civilizations exist somewhere in interstellar space. The failure of most of Sagan's critics to confront this assumption is a sign that they shared his belief in the existence of alien civilizations.

The question was not whether extraterrestrial technological civilizations exist, but what is the best way to communicate with them. And, the most direct method of communication was through the supposed universal medium of science and mathematics. Since the universality of scientific and mathematical knowledge is disputed, all attempts to communicate with extraterrestrials must founder on the fact that they cannot carry their own interpretations with them. They must always suppose the recipient shares the sender's assumptions and thus is able to decode the message.

Pioneer 10 and 11 began their interstellar journeys in 1972 and 1973. By 1977 NASA was ready to launch Voyager 1 and 2 on paths that would also take them past the giant outer planets and beyond the solar system. NASA asked Carl Sagan to supervise the creation of an extraterrestrial message for delivery by the Voyager spacecraft. This time Sagan enlisted a team of scientists and consultants to help him create a means of communication that included photographs and sounds. The multicultural message they composed was so complex and long that Sagan published a 260-page book (*Murmurs of Earth*, 1978) describing its contents and explaining the choice of material.

Working within a six-week deadline, Sagan and his associates devised a gold-coated copper long-playing record. This unique LP contained: 118 photographs of life on Earth; 90 minutes of the world's finest music, including Western classical music, world folk music, and pop hits of the day; greetings to aliens from persons speaking 54 different languages; and an audio essay featuring typical terrestrial sounds—volcanic action, rainfall, frogs croaking, birds singing, tools and machines being used, and the like. Finally, the recorded voices of President Jimmy Carter and Kurt Waldheim, secretary general of the United Nations, addressed any future interceptors of the spacecraft. The President acknowledged divisions among the peoples of the world but looked forward to a time when a united Earth might join "a community of galactic civilizations."

The Voyager message was a terrestrial time capsule intended for study by extraterrestrials in the distant future. It will travel for tens of thousands of years before it comes close to a star. At the launching of Voyager 2 in October 1977, Sagan predicted that billions of years from now, when the Sun was extinguished and the Earth a charred cinder, Voyager would persist intact. Thus, the spacecraft would preserve the last remnants of our civilization.

Technicians encased the Voyager LP in an aluminum cover. The surface of the cover contained etched, coded instructions for playing the record at 16 2/3 rpm using the cartridge and stylus packed nearby in the spacecraft. Representations of the hyperfine transition of neutral hydrogen and the fourteen pulsars used in the Pioneer message reappeared engraved on the record's metal cover. Uranium 238 paint coated the cover so that alien scientists who examined the craft could use the radioactivity of the coating to check the launch time specified by the pulsar "clocks."

Compared to the spare mathematically coded Pioneer message, the Voyager message was pitched to human interests and sensibilities. Sagan and his team responded to local and international politics, current cultural sensibilities, and their belief in the universality of the science and mathematics practiced by humans. A member of Sagan's group remarked that all decisions made about the Voyager message assumed that it had two audiences. The first audience inhabited the Earth; the second lived on planets circling distant stars.

In October 1997, NASA and the European Space Agency launched spacecraft Cassini directed to reach Saturn's moon Titan by the summer of 2004. Following Voyager's example, Cassini carried a message attached to its Titan lander. The message, inscribed on a thin single-crystal diamond disc, is expected to survive billions of years. It consists of 600,000 signatures gathered from 81 countries. Cassini's diamond disc replaced Voyager's outmoded LP, but the intent of the senders was the same: to stimulate public interest in the latest space venture.

NASA scientists continue to follow the course of Pioneer 10. It is over 8 billion miles beyond the Earth and still beams signals to Earth. Pioneer's ability to send signals at a long distance raised a question for space historian Marian Benjamin. If Pioneer 10's primitive technology and low power allows it to send identifiable signals to Earth, why have our radio telescopes failed to detect messages from the superior radio transmitters of advanced extraterrestrial civilizations?

After Contact

Suppose members of an alien civilization retrieved a Pioneer or Voyager spacecraft and decoded its message? What then? What would happen *after* humans contacted an advanced civilization in space?

Scientific speculation about the consequence of contact with extraterrestrials was widely discussed for the first time in the twentieth century. The invention of the radio telescope and the success of space programs in the Soviet Union and the United States popularized the issue of alien contact. Humans could now travel short distances through space, and it was conceivable that aliens with superior technology might visit Earth, send an exploratory space probe, or pass on information in a radio transmission.

In the late nineteenth century, science fiction writers imagined that contact with an extraterrestrial civilization would bring disaster to humanity. By the twentieth century, scientists were more optimistic about the results of alien contact. They stressed the benefits of communication with advanced life in the universe.

H. G. Wells's novel *War of the Worlds* (1898) and Orson Welles's adaption of it as a radio drama (1938) depicted a brutal Martian invasion of Earth. They represent the earlier pessimistic interpretation of extraterrestrial encounters. Carl Sagan, in his popular television series *Cosmos,* claimed that it is pointless to worry about the malevolent intentions of the aliens we contact. Extraterrestrial science and technology far surpass ours and aliens learned long ago how to live peacefully with one another. Moral refinement is one of the by-products of technological advancement.

As early as 1959, physicists Giuseppe Cocconi and Philip Morrison claimed that detection of interstellar signals would have profound practical and philosophical implications for the human race. Consciously, or unconsciously, researchers following in Cocconi and Morrison's footsteps linked the benefits of extraterrestrial contact to problems facing industrial societies during the latter half of the twentieth century. The existing social and political climate influenced what scientists expected to gain from an encounter with intelligent alien life.

In 1961 NASA sponsored an early study of the long-range goals of the space program and its effect upon American life. The participants in the study thought it unlikely that actual meetings with aliens would take place in the next twenty years. In the meantime, NASA should investigate radio contact with intelligent aliens and search for artifacts visiting extraterrestrials may have left behind when they visited the Moon, Mars, or Venus in earlier times.

The study encouraged NASA to explore the emotional, intellectual, social, and political consequences of contact with extraterrestrial civilizations. There was much at stake. Alien contact could end antagonism between rival nations and unify the Earth's population. On the other hand, the unfamiliar ideas and values of advanced extraterrestrial creatures might cripple human society.

NASA astronomer Alastair Cameron edited one of the first anthologies of scientific papers dealing with interstellar communication in 1963. In his introduction to the volume, Cameron endorsed the optimistic scenario of human contact with alien life. He claimed that extraterrestrial knowledge would enormously enrich all aspects of science and the arts and teach us how to establish long-lived world government. Cameron's interpretation of the implications of alien contact typified later scientific discourse on the subject.

The optimistic viewpoint on alien contact prevailed at the first international conference on communication with extraterrestrial civilizations (CETI) held in

Byurakan, Soviet Armenia, 1971. At this meeting, Carl Sagan reminded his audience that any alien civilization we contacted would be superior to ours. He said that alien civilizations have very long lifetimes and extremely advanced technology. By drawing upon their superior knowledge, we may solve the technological problems that plague us in the twentieth century. CETI conference attendees agreed with Sagan and others who claimed that extraterrestrial solutions existed for terrestrial problems.

In the year of the Byurakan conference, NASA sponsored a small-scale study of the feasibility of communicating with intelligent extraterrestrials. This study, proposed by John Billingham of NASA's Ames Research Center, was named Project Cyclops. It called for the deployment of an array of 1,000 to 2,500 connected radio dish antennae aligned to search for extraterrestrial intelligent life. Project Cyclops was never implemented, but its participants recommended that NASA make the search for extraterrestrial intelligence an ongoing part of the NASA space program.

The project's final report also included a discussion of the human implications of alien contact. Bernard Oliver, who wrote the report, listed and then dismissed the possible hazards of contact with intelligent extraterrestrials. Instead, he stressed the many benefits to science and technology that awaited the recipients of a message from an alien civilization that had long outgrown its technological infancy.

Once we establish contact with the extraterrestrials, Oliver wrote, we will have access to the enormous body of galactic knowledge assembled over eons of time. This treasury of cosmic knowledge will reveal the secrets of the universe, including its origins and ultimate fate, and allow us to identify ourselves with a supersociety. Surely, he concluded, such information alone was worth several times the cost of Cyclops, estimated at $6 to $10 billion over 10 to 15 years.

Sagan directly addressed the technological problems facing modern society in his book *The Dragons of Eden* (1977). He listed overpopulation, disparity between rich and poor nations, shortages of food and natural resources, environmental pollution, and the threat of a nuclear holocaust. Then he wrote that the first extraterrestrial message we intercept might contain detailed instructions on how to avoid technological disasters and achieve stability and longevity for our species. Sagan concluded that the relatively inexpensive search for signals from extraterrestrial civilizations promised much for the future of the human race. No comparable enterprise could match it.

By the early 1970s, searchers for extraterrestrial intelligence adopted two basic premises. First, the establishment of communication with an alien civilization would affect scientific, technological, and philosophical thought on Earth and lead to positive social and political changes. Second, failure to establish contact would reveal the unique position humans occupy in the universe and force them to rethink the meaning of terrestrial life. Win or lose, the search for extraterrestrial intelligence would change the ways humans perceive themselves and the universe. As Sagan said, the search for intelligent aliens is unique. It is one of the few human endeavors where failure is counted as success.

Sagan's biographer, Keay Davidson, criticized the notion that answers to problems we face on Earth lay hidden in messages coming from the heavens. Davidson recounted the state of American society in the 1970s: the divisive war in Vietnam, crime and unrest in the cities, the growing use of illicit drugs, and the energy crisis. In those troubled times, Davidson remarked, some of our best scientific minds had nothing better to offer than salvation from the stars.

Davidson noted that those who advocated the search for signs of extraterrestrial intelligence unwittingly projected technical assistance given by Western nations to Third World countries into the wider universe. They assumed that advanced alien civilizations would provide us with technical help just as contemporary developed nations helped the underdeveloped countries of the world. Sagan and like-minded scientists expected extraterrestrial creatures to send them a quick technological fix for the serious historical, social, and economic problems Western nations faced in the 1970s.

Not all scientists agreed with Sagan and his colleagues about the implications of the first human encounter with extraterrestrial intelligence. Nobel prize winning biologist George Wald expressed his reservations at a symposium sponsored by Boston University and NASA in 1972. He feared what might happen if humans contacted aliens who possessed superior technology. It "does not thrill me,"[11] he said, to be dependent upon the advanced science and technology of an alien civilization. Extraterrestrial communication might excite a handful of researchers, Wald added, but it would endanger the lives of the rest of us.

Wald acknowledged that humans faced crucial problems, such as finding a cure for cancer and controlling thermonuclear reactions, but he insisted that humans solve their problems without seeking help from outer space. The entire human enterprise might collapse if we relied upon aid provided by advanced alien civilizations. Although Wald spoke forcefully, he converted few of the main participants in the symposium. Sagan and several physical scientists disagreed with him, as did anthropologist Ashley Montagu and theologian Krister Stendahl.

A few years later Sagan and Drake discovered that simply sending a message to alien civilizations was a controversial act. In November 1974, Frank Drake and the staff at the National Astronomy and Ionosphere Center used the 1,000-foot-diameter radio telescope at Arecibo, Puerto Rico, to deliver a powerful coded radio signal to 300,000 stars in the globular cluster in the constellation Hercules. Although NASA and the National Science Foundation funded the telescope and its radar transmitter, the Drake/Sagan transmission was not a NASA-sponsored event. It was part of the ceremony to mark the reopening of the observatory after a recent upgrade of its equipment.

Sagan helped Drake devise the coded message for transmission to the globular cluster of stars. The message consisted of 1679 binary digits expressed as 0 or 1. Savvy extraterrestrials were expected to know that 1679 is the product of two prime numbers, 23 and 73. When the digits are arranged in 73 rows of 23 characters each, and 1 represented as a black filled square and 0 as an empty white one, a visual message emerges in the pattern of checkered squares (Fig. 7.2).

The visual message began with the numbers one to ten in binary notation and then moved on to the atomic numbers of key chemical elements and to the structure of DNA. Immediately below the DNA helix, Drake placed a schematic drawing of a human figure, a sketch of the solar system, and a simplified image of a radio telescope. Thus, the Arecibo team assumed that aliens who received their message understood terrestrial mathematics, chemistry, and biology, as well as the technology of radio telescopes.

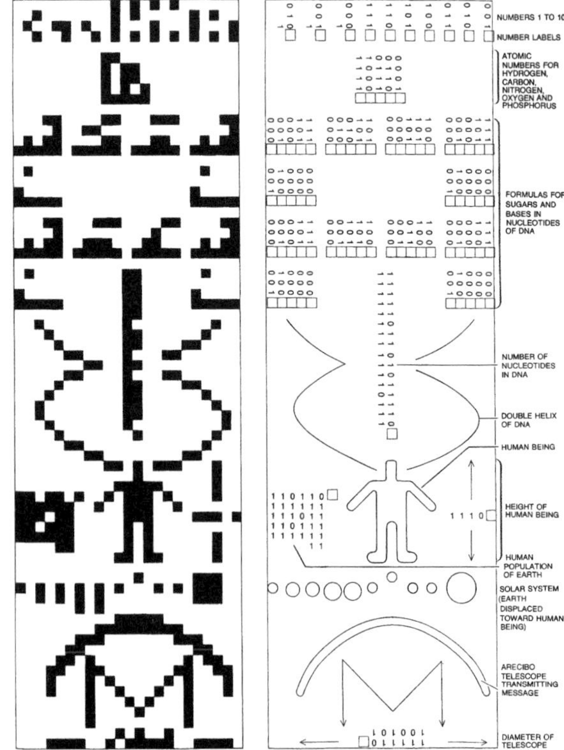

FIG. 7.2. Representation of the 1974 Arecibo message, the first digital, encoded radio message transmitted to outer space. Left panel: message transcribed into black and white squares. Right panel: explanation of information contained in left panel. (From "Search for Extraterrestrial Intelligence," Carl Sagan and Frank Drake. *Scientific American*, May 1975, p. 87. Copyright 1975 by Scientific American, Inc. All rights reserved.)

The prominent English astronomer Sir Martin Ryle reacted immediately to Drake's interstellar transmission. Ryle brought a formal complaint to the International Astronomical Union. He asked the union to halt all further attempts by astronomers to contact other civilizations because of the possible harmful consequences to the human race. Ryle worried that terrestrial contact with intelligent aliens might lead them to invade the Earth with the intention of colonizing us or stealing our mineral resources.

The *New York Times* (Nov. 22, 1976) responded to Ryle with an editorial entitled "Should Mankind Hide?" The *Times* editorial writers argued that any distant civilization would be superior to ours and have no need to use the crude methods of domination employed by Columbus in the New World. On the contrary, intelligent aliens might offer us a cure for cancer or the knowledge for controlling thermonuclear energy.

Some commentators pointed out that high-powered radar and television transmitters sent signals into space decades before the Arecibo team tried to communicate with a coded radio message. Writing in the June 1974 issue of *TV Guide,* Carl Sagan pointed out that humans unintentionally had been broadcasting radio and television messages into space since the early years of the twentieth century.

The many benefits to humanity promised by SETI promoters made them vulnerable to criticism from skeptics. In a 1985 essay debunking the search for extraterrestrial intelligence, philosopher Edward Regis, Jr., attacked the claim that failure to find any evidence of it would influence the future of humankind. He cited Carl Sagan's assertion that the discovery that humans are alone in the universe would bring together hostile nations on Earth.

Regis pointed out that for centuries before Copernicus, humans fought disastrous wars despite their belief that they lived alone in the universe. Why should the modern rediscovery of human uniqueness bring peace to the Earth? Why should it differ from the situation that existed before the Copernican principle of mediocrity suggested that intelligent creatures populate remote planets?

Regis also said that it is nearly impossible to prove that humans are alone in the universe. Searchers for extraterrestrial intelligence were fond of claiming that absence of evidence is not evidence of absence. It is always possible to dismiss any search project as flawed and propose a new search for alien life based on a different set of assumptions. This is what happened during the century-long debate over life on Mars. Believers in Martian life persistently explained away failure and called for another round of observations or tests.

Regis doubted that any test could prove conclusively that humans are the sole intelligent creatures in the cosmos. Neither was he impressed by the promise of SETI supporters that contact with advanced civilizations would yield dividends for the human race. He reasoned that extraterrestrial knowledge would come from creatures wholly unlike ourselves who live in vastly different physical and cultural environments. Hence, there may be nothing useful for the human enterprise in an extraterrestrial message. Regis was reluctant to justify the search for alien intelligence on the grounds of its large-scale benefits for humanity.

The End of the Beginning

NASA's interest in the search for extraterrestrial intelligence began modestly. It was advanced by individual researchers and not by overt agency policy. During the 1970s, SETI at NASA took the form of workshops, conferences, and reports on the topic. By 1979 NASA was ready to embark on a ten-year funded program that would culminate in a full-fledged search for radio signals from advanced extraterrestrial civilization. This search officially opened in October 1992.

In 1991, as NASA readied its equipment for the coming search for extraterrestrial intelligence, it sponsored a series of workshops on the social, cultural, and political consequences of the detection of an extraterrestrial civilization. Held in October 1991 and May and September 1992, the workshops drew a mixed group of participants. It included academic specialists in the social and behavioral sciences, astronomers, journalists, educators, and assorted NASA personnel.

NASA asked this interdisciplinary group to assess the short- and long-term responses of societies to the discovery of alien intelligence, advise the space agency how to inform the public about its discoveries, and recommend areas for additional study. NASA was preparing itself for positive results from its official entry into the search for extraterrestrial intelligence.

The published proceedings of the workshops are conservative in their recommendations and forecasts. At times, however, remnants of the older optimistic outlook surface in its pages. For example, the authors of the section on history and SETI carefully review the historical context of the search for alien intelligence. However, as a coda to their report, they print several lines from a 1968 science fiction novel, *The Cassiopeia Affair.* The quoted material describes an initial terrestrial encounter with intelligent aliens: "We believe that Cassiopeian civilization is technologically more advanced than our own [and that they] might well provide us a sort of cosmic technology assistance program for this emerging Earth."[12]

Throughout their various reports, the workshop members emphasize how difficult it is to forecast consequences of alien contact. Their recommendations range from suggestions that NASA target influential groups to disseminate information on SETI research to concerns that international organizations receive prompt and proper news about any alien signals.

When NASA's survey of the skies for alien radio signals formally began in the fall of 1992, a new round of speculation about the consequences of alien contact developed. But it began too late. NASA's much publicized search ended within a year. A message arrived at NASA headquarters, not from the stars but from members of Congress. Unmoved by promises of the benefits of alien contact, they halted funding for NASA's SETI program. The search for alien civilizations was now left to privately funded research institutes as well as astronomers at private and public universities.

CHAPTER EIGHT

Life in an Expanding Universe

> SETI is often compared to a religion . . . by its critics. But, in fairness, the resemblance holds, largely because of the seeking and searching that underpins the enterprise, technically and metaphorically. Unlike other scientific experiments, SETI searches do not yield conclusions (at least they haven't so far). They just elaborate on methodology. At best, they suggest that a particular star or a particular part of the sky may be discounted as a rich hunting ground for alien intelligence—and then, only discounted using this very specific and narrow method for divining its existence.
>
> —Marina Benjamin, *Rocket Dreams*, 2003

Skepticism and Acceptance

During the first half of the twentieth century, many astronomers were skeptical about the existence of life beyond our solar system because they believed that planetary systems were not common in interstellar space. In 1923 English astronomer Sir James Jeans (1877–1946) lectured that the Earth alone supported life, although he admitted it might exist elsewhere. Six years later, another English astronomer, Arthur S. Eddington (1882–1944), argued that nature created the universe solely to "provide a home for her greatest experiment, Man."[1]

The Astronomer Royal Sir Harold Spencer Jones (1890–1960) offered a different assessment of the situation in his popular book, *Life on Other Worlds* (1940). He accepted the formation of planets around stars as normal events in stellar evolution. If planetary systems are common in the universe, he wrote, then it is likely there are Earthlike planets supporting life.

The mid-twentieth-century shift toward the acceptance of extraterrestrial life appears in the writings of yet another eminent English scientist, the astrophysicist and cosmologist E. A. Milne (1896–1950). Early in his career, Milne studied at Trinity College, Cambridge, with Ernest W. Barnes. In the 1930s, Barnes first raised the possibility of using radio receivers to detect signals sent by intelligent extraterrestrials.

Shortly before his death, Milne delivered a series of lectures on modern cosmology and the Christian idea of God. Milne repeated Eddington's observation that just as a bountiful nature scatters millions of acorns to grow one oak tree, so does she create an infinite number of star-filled galaxies to grow one planet supporting life.

Milne, however, interpreted Eddington's observation differently. Consider, he said, that the infinity of galaxies are the scenes of an infinite number of experiments in biological evolution. The ordinary operation of Darwinian natural selection and Mendelian genetics affecting life on an infinite number of planets produces an infinite variety of living things. Milne, a chaired professor of mathematics at Oxford University, then turned to a line of argument first used in the Middle Ages.

Is it irreverent to suggest, Milne continued, that an infinite God could neither enjoy Himself nor properly exercise His powers "if a single planet were the sole seat of His activities?" "God . . . did not wind up the world and leave it to itself,"[2] Milne asserted. The Creator guided the subsequent evolution of life an infinite number of times in an infinite number of places.

Milne filled the universe with a diversity of life. Now he faced the question that troubled earlier Christian thinkers who proposed the existence of intelligent life on other worlds. Were the crucifixion and resurrection of Jesus Christ unique terrestrial events, or did Christ repeat them on countless other planets? Milne answered that a Christian could never accept the suffering of the Son of God on planet after planet. His response to the possibility of Christ's incarnation on other worlds was influenced by ideas from the late Middle Ages and recent advances in science and technology.

The fifteenth-century theologian William Vorilong claimed that Christ's crucifixion simultaneously redeemed all intelligent creatures in an infinite universe. Milne proposed that news of Christ's Atonement on Earth traveled swiftly to all parts of the universe. Astronomers using radio telescopes, he reported, had recently discovered signals emanating from sources in the Milky Way. It was possible that the signals came from intelligent beings on other planets.

Milne suggested that an interstellar communication system using radio signals operated two thousand years ago. Intelligent creatures used this vast radio network to broadcast news of Christ's sacrifice for humanity to the entire universe. The transmission of signals by radio to other planets, Milne argued, made it unnecessary to reenact the crucifixion endlessly throughout the cosmos.

Milne's treatment of Christ's incarnation in an infinite universe demonstrates the power and longevity of medieval religious ideas and calls attention to recent scientific developments that led Milne and others to reconsider the possible existence of extraterrestrial life. Milne mentioned radio astronomy, and he drew upon new hypotheses about planet formation in an expanding universe.

The idea of an expanding universe originated in the first half of the twentieth century, but its observational basis began a century earlier. Stellar astronomy developed in the nineteenth century in conjunction with an increase in telescope size. Thereafter, European and American telescope builders competed to construct instruments with ever larger reflecting mirrors.

The new surge of interest in stellar astronomy and large expensive telescopes bypassed Percival Lowell. He could not match the wealth of the philanthropists who funded large-mirror American telescopes, nor was he attracted to stellar astronomy. Planetary astronomy suited Lowell's pocketbook and was in line with his fascination with Mars. Personnel at the Lowell Observatory, however, made preliminary observations crucial to stellar astronomy and the idea of an expanding universe.

In 1909 Lowell directed observatory staff member Vesto M. Slipher to study the spectrum of the spiral nebula Andromeda. Slipher's spectrographic analysis of Andromeda showed a displacement of lines toward the violet. He determined that the displacement meant the Andromeda nebula was traveling at a great velocity—more than 300 kilometers per second. Lowell urged Slipher to study other spiral nebulae to confirm his findings.

Slipher's discovery proved to be one of the greatest in recent times. It supported the idea that spiral nebulae were not nebulous stars or solar systems in the early stages of development. They were gigantic star-filled galaxies located at enormous distances from the Milky Way. Slipher's researches on spiral nebulae 1914–1917 motivated astronomers to gather data on the velocities of a number of spiral nebulae, or as they came to be called, spiral galaxies. The Milky Way and Andromeda are now classified as spiral galaxies.

By the early 1930s, evidence accumulated about galaxies was ready for interpretation. Drawing upon the research of astronomers and physicists, the astronomer Edwin P. Hubble (1889–1953) developed the idea of an expanding universe. He determined that the velocity of a receding galaxy is proportional to its distance from an observer. The greater the velocity, the more distant the galaxy. All galaxies appeared to be moving away from the Milky Way in accordance with this law. Hence, the universe was not at a standstill. It was expanding at a rapid rate with the galaxies receding from one another. Hubble's conception of an expanding universe filled with galaxies containing billions of stars led him to conclude that life-supporting planets existed among the stars.

Origins of Life

The discovery of new galaxies, and the observational proof that the universe as a whole was expanding, forced astronomers to rethink the issue of extrasolar planets. During the fifteen-year period 1943–1958, astronomers became convinced that planetary systems and inhabited planets were not rare occurrences.

Along with an expanding universe, new scientific studies of the origins of life influenced the acceptance of extraterrestrial life at midcentury. Questions about the origins of life moved from the realms of philosophical theorizing to the laboratories of chemists. Once scientists began investigating the chemical origins of terrestrial life, they found it easier to imagine similar processes taking place on other worlds.

Chemist Melvin Calvin (1911–1979) was one of the first scientists to carry out experimental investigations on the origins of life. In 1950 he irradiated a mixture of water vapor and carbon dioxide with high-energy alpha particle radiation from a cyclotron. Calvin did not create living matter in this series of experiments. He did prove that given proper conditions organic compounds can arise from inorganic substances. Calvin's investigations inspired other chemists to follow his lead.

In 1953 Stanley Miller demonstrated that conditions thought to exist on the primitive Earth could produce the amino acids essential to life. Within a year, Harvard biologist George Wald extended the new chemical understanding of life to the universe. He wrote: "Wherever life is possible, given time, it should arise."[3] Wald estimated that there were 100,000 Earthlike planets in our Galaxy, each one a potential site for life.

Melvin Calvin was a brilliant chemist who considered the origins of life on Earth and elsewhere in the universe. In the 1950s, he influenced a trend in thinking about extraterrestrial life and intelligence that became popular among scientists and the public in the latter half of the twentieth century. Calvin envisioned evolutionary forces working at every level, from the creation of the first complex chemical compounds to the establishment of intelligent life throughout the universe.

Calvin drew upon modern chemistry to develop a theory of chemical evolution. He also borrowed the concepts of random variations and selection from biological evolution and applied them to chemical systems. Chemists have already demonstrated, he said, that physical conditions on the primitive Earth transformed a random mixture of inorganic substances into organic ones. Given the right energy sources, temperatures, and physical environments, evolving organic molecules eventually acquired the attributes of living material. Once life appeared, Darwinian biological evolution supplanted chemical evolution, and the wonderful variety of living organisms emerged on Earth. Humanity, Calvin declared, is the most highly developed form of organic evolution.

Calvin next extended evolution beyond our planet. Drawing upon the optimistic predictions of planetary systems made by astronomers, Calvin postulated the existence of millions of Earthlike planets in the universe. He believed many of these planets duplicated the evolutionary processes operating on Earth.

Calvin speculated that some planets around other stars support intelligent life forms much older than the human race. These extraterrestrials, he concluded, may be "far more skillful and knowledgeable than we"[4] in matters of communication. Therefore, astronomers should use radio telescopes to search for messages sent by advanced organisms living beyond the solar system.

Calvin's explanation of the origins of intelligent life in the universe influences our thinking about the subject today. Many contemporary commentators on life in the cosmos see continuity extending from chemical to biological evolution, from biological evolution to social and cultural evolution, and from terrestrial to extraterrestrial life. This grand view of the universe is awe inspiring but flawed. Since scientists have not yet created life in their laboratories, the transition from chemical to organic evolution is not assured. Furthermore, the transfer of evolutionary concepts like random variations and natural selection to human cultures is filled with problems. Evolution used in these instances serves metaphorical or analogical purposes but does not carry the certainty of scientific knowledge.

Panspermia

Although scientists have learned much about how life may have developed from complex molecules on the primitive Earth, their work has barely begun. In 1966 Carl Sagan predicted that scientists would synthesize simple life forms in the laboratory within a decade. Thirty-three years later, a noted British paleobiologist, Simon Conway Morris, observed that despite decades of research into the origins of life, and frequent claims of a breakthrough, "we are still paddling on the edges of an ocean of ignorance."[5]

Some researchers believe that life may not have originated on Earth at all. They argue that life arrived here from outer space. In the nineteenth century, a number of prominent physicists and chemists endorsed the idea of panspermia (seeds everywhere). They claimed that life, like matter, was eternal but that the former did not arise from the latter on our planet. Instead, life came to our planet as heat-resistant bacterial spores carried by meteorites, or some other means. Enthusiasm for panspermia declined in the early twentieth century when experiments suggested that living spores could not survive exposure to radiation in space.

Nobel laureate Francis Crick and chemist Leslie Orgel revived panspermia in 1973. They proposed a theory of directed panspermia. According to them, intelligent aliens sent living microorganisms to Earth aboard a special long-range robotic spaceship. The authors speculated that the extraterrestrials intended either to demonstrate their technological prowess or deliberately spread life throughout space.

Crick and Orgel presented directed panspermia as a viable scientific alternative to biochemical theories of the origins of life on Earth. They noted that all terrestrial life shares the same genetic code. This meant that either life encountered an early evolutionary bottleneck that left behind a small interbreeding population, or that life evolved from organisms sent from another planet. Crick admitted that directed panspermia was a premature theory whose time might never come. Nevertheless, he accepted its scientific validity.

Radio Astronomy

Once radio astronomy was established, it became another factor influencing mid-century thinking about extraterrestrial life. Ernest W. Barnes (1874–1953), mathematician, Anglican bishop, and teacher of E. A. Milne, was one of the earliest scientists to propose radio communication with intelligent extraterrestrial beings. In a paper delivered before the British Association for the Advancement of Science in 1931, Bishop Barnes claimed that alien "beings exist who are immeasurably beyond our mental level."[6] These creatures communicate with Earth by transmitting coded radio signals to us. The reception and decoding of these communications, he declared, would inaugurate a new era in the history of humanity. Bishop Barnes spoke shortly before the era of radio telescopes made it possible to receive radio signals transmitted from deep space.

Radio astronomy is the study of the radio waves naturally emitted by bodies in space. In 1928 the Bell Telephone Laboratories in New York City hired physicist Karl G. Jansky (1905–1950) to investigate static disrupting radiotelephone transmissions. By the early 1930s, Jansky isolated a distinctive and persistent static hiss coming from the center of the Milky Way. The *New York Times* featured Jansky's discovery on its front page on May 5, 1933, reporting that the incoming signal was natural. The paper made it clear that there was no evidence "of interstellar signaling."

Jansky's discovery stirred little interest among astronomers accustomed to using light-gathering telescopes. Bell Laboratories assigned Jansky other research topics, and interest in the detection of radio signals from space languished until the Second World War. The development of military radar stimulated scientists and engineers to reexamine Jansky's work after the war ended.

Before the establishment of radio astronomy, observation of the heavens was limited to the visible portion of the electromagnetic spectrum. The invention of radio telescopes made it possible to detect celestial objects that radiated electromagnetic waves with wave lengths beyond that of light. Radio telescopes, for instance, detected hitherto unknown radio signals emitted by the Sun. These solar observations facilitated a new understanding of the physics of the Sun. Radio telescopes were used in the discovery of pulsars, rapidly spinning neutron stars that emit pulses of radiation, and quasars, very bright, distant stars that are sources of radio signals. Finally, radio astronomy provided evidence for the Big Bang theory of the universe by identifying residual background radiation left over from the initial Big Bang.

Some scientists realized that radio astronomy technology was also useful in searching for interstellar communication by intelligent beings. The first full-fledged scientific statement on the search for interstellar communication appeared in 1959 when physicist Giuseppe Cocconi (1914–) and astrophysicist Philip Morrison (1915–) published a short article outlining the physical parameters of communications with advanced alien civilizations.

The authors settled on the radio band for communication purposes. They chose the hyperfine radio emission line of neutral hydrogen, whose wavelength was twenty-one centimeters. Aliens were likely to communicate at that wavelength because cosmic interference was minimal and hydrogen was an abundant element in the universe. Any advanced civilization was assumed to know the physical characteristics of the hydrogen atom.

Cocconi and Morrison realized the highly speculative nature of their proposal. Hence, they concluded their paper modestly. "The probability of success is difficult to estimate"; they wrote, "but if we never search, the chance of success is zero."[7]

The Cocconi-Morrison paper became a classic document in the search for extraterrestrial intelligence and stimulated other researchers to test the authors' hypotheses. Cocconi soon returned to work in high energy physics, but Morrison retained a life-long interest in these matters. He was a pioneer in the scientific search for intelligent life on other worlds.

While Cocconi and Morrison were writing their paper, astronomer Frank Drake was readying the eighty-five-foot diameter radio telescope of the National Radio Astronomy Observatory (NRAO) at Green Bank, West Virginia (Fig. 8.1). He intended to use it for the reception of signals transmitted by intelligent extraterrestrials. Drake independently selected the hyperfine radio emission line of neutral hydrogen as the wavelength extraterrestrials would likely use to transmit interstellar messages.

The choice of the same wavelength by Cocconi, Morrison, and Drake was not coincidental. In 1951 physicist Edward M. Purcell (1912–1997) discovered that hydrogen clouds in space emitted a signature wavelength of twenty-one centimeters. Purcell's discovery permitted radio astronomers to chart the movement of hydrogen clouds in space. Those who used the twenty-one-centimeter wavelength for message reception assumed that extraterrestrials practice physics as we do and arrive at the same conclusions about the physical world.

FIG. 8.1. First major radio telescope erected at the National Radio Astronomy Observatory, Green Bank, West Virginia. (Courtesy of National Radio Astronomy Observatory/AUI.)

Drake thought that the Green Bank telescope was capable of receiving messages from stars located within ten light years of the Earth. He proposed minor modifications to the instrument that would enable it to detect incoming radio signals at the twenty-one-centimeter wavelength. Drake received permission to use the telescope as a detector of artificially generated interstellar signals.

Drake chose two nearby stars for his initial study, Tau Ceti and Epsilon Eridani. On April 8, 1960, Drake's team aimed their telescope at Epsilon Eridani and recorded a very strong incoming signal. They soon learned that the signal was due to terrestrial interference, probably from nearby aircraft. Drake spent 200 hours examining radio signals from the two stars without discovering any signs of alien communication with the Earth.

The year 1959–1960 marked the beginning of a new era in the search for extraterrestrial intelligence. Cocconi and Morrison provided a reasonable theoretical basis for the search, and Drake showed how to conduct the search with existing equipment. The accomplishments of this pioneering trio encouraged others to join the search.

Drake's Equation

The Space Science Board of the prestigious National Academy of Sciences responded positively to recent advances in the study of extraterrestrial life. It authorized a small conference on the subject at the National Radio Astronomy Observatory located at Green Bank, West Virginia. Notable scientists attended this meeting in November 1961. Attendees included Melvin Calvin, Giuseppe Cocconi, Frank Drake, John C. Lilly, Philip Morrison, Bernard Oliver, and Carl Sagan.

John Lilly, a popular researcher of dolphin intelligence, opened the Green Bank conference. He spoke about his research on communication with dolphins. Dolphins have brains slightly larger than humans and a density of brain nerve cells similar to a human brain. These facts convinced Lilly that dolphins possess intelligence comparable to humans. Lilly believed that dolphins had developed a complex language—he called it dolphinese—and that he would eventually decipher it. He predicted that by the 1980s, humans would establish communication with another species, if not extraterrestrial, then a marine organism on Earth.

Several of the participants drew parallels between dolphins and extraterrestrials. If dolphins are as intelligent as humans, then intelligence has emerged independently more than once on Earth. This was proof that intelligence was not a rare commodity in the universe. Scientists who learned how to communicate with intelligent dolphins could develop techniques to communicate with intelligent aliens.

Lilly's fellow conferees realized that there were drawbacks to the comparisons he drew between dolphins and intelligent extraterrestrial life. Dolphins spend all of their time in water, they have no hands, and their superior intelligence did not lead them to develop technology. Similarly, extraterrestrials confined to an aquatic planet might be intelligent but incapable of using fire or building a radio transmitter to send messages to Earth.

Frank Drake, who was responsible for the content of the Green Bank meeting, looked for a general principle to focus the discussion. In the process, he produced an equation that included the key factors needed to determine the probable number of intelligent communicative civilizations present in the Galaxy. His pioneering effort is called the Drake or Green Bank equation.

Drake's equation was the first attempt to quantify the *probability* of communicating with an extraterrestrial civilization. It became the single most important formulation in the search for extraterrestrial intelligence and has remained popular for over forty years. Drake's equation was not entirely original. Several of his contemporaries had made similar calculations.

Drake's use of probability in his equation needs clarification. The determination of the probability of finding extraterrestrial life on an extrasolar planet should not be confused with the *relative frequency* approach to probability. In the latter case, one determines the probability of a flipped coin showing heads or of a given card being picked from a deck of cards.

The approach used in Drake's equation falls under the rubric of *personal* or *subjective* probability. Here the probability assigned to an outcome depends upon the subjective judgment of the investigator. Two persons with different knowledge, experiences, and feelings might assign different probabilities to the same event.

Drake incorporated subjective probability in a comprehensive equation he wrote for the determination of advanced life on other worlds. His equation was presented in a simple mathematical form. Its interpretation and solution, however, were never simple. They created problems and controversies that remain unresolved today. Drake's equation reads as follows:

$$N = R_\star \ f_p \ n_e \ f_l \ f_i \ f_c \ L$$

N is the number of civilizations in our Galaxy with the technological competence to transmit and receive radio signals. If N is very small, then we are unlikely to detect any incoming signals. If it is large, a search that begins with nearby stars is a worthwhile strategy because of a reasonable probability of finding civilizations close to us.

The value of N is determined by multiplying the terms on the right side of the equation. These terms represent factors influencing the existence of intelligent communicative life in the universe.

$R_\star$ mean rate of star formation in the Galaxy.
f_p fraction of stars having planets.
n_e number of planets per star with environments capable of supporting life.
f_l fraction of habitable planets that actually support life.
f_i fraction of life-bearing planets on which intelligent life appears.
f_c fraction of intelligent civilizations that might communicate with the rest of the Galaxy.
L lifetime (in years) of technically advanced civilizations.

To summarize: We need to know the mean rate of star formation in our Galaxy when the solar system emerged, the fraction of those stars having planets, the fraction of the above planets that can support life that actually do, the fraction of life-supporting planets that have developed intelligent life, the fraction of planets that have the means and are willing to communicate using radio waves, and the average lifespan of extraterrestrial technological civilizations.

The first three factors are astronomical, the next two biological, and the last two social. When Drake wrote his equation on the blackboard for the Green Bank audience, no astronomer could state with certainty that any star, other than the Sun, had a planetary system.

Although there were physicists, astronomers, and biologists at the Green Bank meeting, no one represented the social sciences. There was no one competent to discuss the nature of civilizations, the character and relative proliferation of technological civilizations, and the longevity of technological civilizations.

The strength of Drake's equation is its simplicity and its combination of astronomical, biological, and social factors. At a critical time in the search for extraterrestrial life, the equation neatly summarized the outstanding issues for discussion. Drake's equation does not make a fundamental statement about the nature of the physical world. Graham Farmelo, who listed the Green Bank formulation among the greatest equations of science, said that Drake "put the subject into a truly scientific-looking format."[8]

Estimation of the first factor in Drake's equation, the mean rate of new star formation in our Galaxy, came from the astronomers who met at Green Bank. They guessed conservatively that one new star would form each year but were willing to consider as many as ten. The solution of the first factor was easier than judging the number of stars in our Galaxy with planetary systems (factor two). At that date, astronomers knew only one such star, our Sun. The Green Bank estimate was that 1/5 to 1/2 of all stars possessed planetary systems. When the discussants weighed the importance of the remaining factors, the estimates became more speculative.

Astronomers next estimated the third factor, the number of planets per star with life-supporting environments. Using our solar system as their guide, they estimated one to five planets. Factor four was the fraction of planets with environments suitable for life that supported life. Sagan and Calvin gave an optimistic response. They claimed that given enough time, life would eventually appear in a suitable environment.

Lilly and Morrison joined forces in assessing the fraction of intelligent life that would appear on life-bearing planets (factor five). They argued that intelligence favored survival of organisms and that intelligence would inevitably arise along the path of organic evolution. Morrison reminded the group of the importance of convergence in evolution, the tendency for different species to adapt in similar ways to environmental demands. Morrison cited Lilly's recent work on dolphins to prove that two radically different organisms on Earth had converged on intelligence. The conferees concluded that wherever there was life in the universe, it would show signs of intelligence.

By this time the Green Bank group had reached the last two factors, numbers six and seven. The evaluation of these factors needed the help of historians, anthropologists, and sociologists, who were not at the meeting.

The sixth factor was the fraction of intelligent extraterrestrial societies to develop the technology and interest to communicate across deep space. The Green Bank participants scavenged their knowledge of world history seeking an answer to the questions posed here. How could one be certain that every alien civilization would develop communication technology? If an advanced civilization devised the appropriate communication technology, would they necessarily use it? After much inconclusive discussion, the participants decided that from one-tenth to one-fifth of intelligent extraterrestrial societies might attempt to signal other parts of the Galaxy.

Factor seven, the lifetime of the technologically advanced communicative societies, was the most important one. Some members of the Green Bank assembly, taking their cue from the ongoing Cold War, envisioned technological civilizations destroyed in nuclear holocausts. Others thought that advanced civilizations might exhaust their natural resources, as industrial societies were doing on Earth, and expire. Still others imagined epidemics or internal structural collapse ending an era of spectacular technological growth. The participants decided after much conflicting discussion that the lifetime of advanced technological civilizations ranged between one thousand years and more than one hundred million years. If a technological civilization managed to escape early destruction, then its chances of long-term survival were excellent.

Multiplying Drake's factors yielded an answer for the critical number N. The result was that between one thousand and one billion Galactic civilizations possessed the capability and interest to communicate with the Earth. Most of the attendees felt the higher number was more accurate.

The Green Bank conference accomplished much more than offering tentative answers to the questions posed by Drake's equation. It gave legitimacy to the search for extraterrestrial intelligence. Carl Sagan remembered the Green Bank meeting fondly as a time when a small group of reputable scientists gathered to discuss intelligent life on other worlds.

Historian of astronomy Steven J. Dick assessed the Green Bank results somewhat differently. He declared: "Perhaps never in the history of science has an equation been devised yielding values differing by eight orders of magnitude. . . . each scientist seems to bring his own prejudices and assumptions to the problem."[9]

A few scientists judged the Drake equation useless. At one point, Joshua Lederberg called it "Hocus-pocus,"[10] and SETI researcher Christopher Chyba warned that it "is a shorthand for what we are trying to understand, rather than a tool for precise calculation."[11] Others attempted to modify features of the original equation. Nevertheless, at the beginning of the twenty-first century, Drake's formulation continues to exert its influence on thinking about intelligent alien life. In 1998 science writer Amir D. Aczel devoted an entire book to the resolution of the equation. Using modern probability theory, Aczel concluded that life exists on at least one other planet somewhere in the universe.

• • •

Shortly after the Green Bank meeting, Frank Drake sent the participants a hypothetical interstellar message. Drake intended to show the form such a message might take and the information it might convey. He invited his Green Bank colleagues to try their hand at deciphering his contrived message from the stars.

Drake's message consisted of 551 ones and zeros arranged consecutively as they might appear on a continuous piece of tape. No single recipient deciphered the entire message, but different individuals solved various portions of it. Drake translated his message into the purported universal language of mathematics. Mathematicians know that 551 factors into two prime numbers, 19 and 29. This suggests arraying the ones and zeros into 29 groups of 19 characters each, or vice versa. Neither of these choices makes any sense until the recipient of the message converts the ones and zeros into squares of two colors, say black and white. When correctly assembled, the result resembles an unsolved crossword puzzle.

Crude, but recognizable, pictures emerge when the recipient arranges the contrasting squares into twenty-nine groups of nineteen squares each. In the bottom center of the array of squares is a primate-looking figure that, according to the coded information, stands about ten feet tall. The figure's two legs are planted on the ground, his head is in the air, and his arms are stretched out along his torso. Along the left margin, Drake displayed a sketch of the alien's solar system with its nine planets. The schematic drawings of the oxygen and carbon molecules at the top right of the array indicate the messenger's life chemistry.

Planet 4, the home planet of the message transmitter, has a population of 7 billion inhabitants, a number provided in binary notation. Since planet 3 has a small population of 3,000, it must be a colony of planet 4. The eleven inhabitants of planet 2 represent a team of scientists exploring the body. Obviously, Drake added these fictional details to make the message more interesting and challenging.

Like Drake's equation, his interstellar message appeared in books and articles written by scientists and popularizers of science. Melvin Calvin included it at the end of an article he wrote for *Science Digest* (1963) and promised a solution in a later issue of the magazine. Science journalist Walter Sullivan reprinted it in his very influential book *We Are Not Alone: The Search for Intelligent Life on Other Worlds* (1964). Sagan and Shklovskii discussed it at length in *Intelligent Life in the Universe* (1966).

Drake's squat little figure, fashioned from black squares, became as popular as his equation. The reproduction of Drake's message in books written by well-known scientists has given it an aura of authenticity approaching that of an actual communication from the stars.

Drake's interstellar message has attained a longevity and authenticity that its author probably never intended. It gained special appeal from its association with the binary code used in electronic computing. However, the form of Drake's message was not new.

Forty years earlier, two contributors to *Scientific American* (March 20, 1920) proposed a similar system for communicating with intelligent aliens. In an article entitled "What Shall We Say to Mars?" H. W. Nieman and C. W. Nieman proposed using Morse code to establish contact with Martians (Fig. 8.2).

The Niemans did not intend to teach Martians the basics of Morse code. They recommended sending dots and dashes, as flashes of light, with a long pause

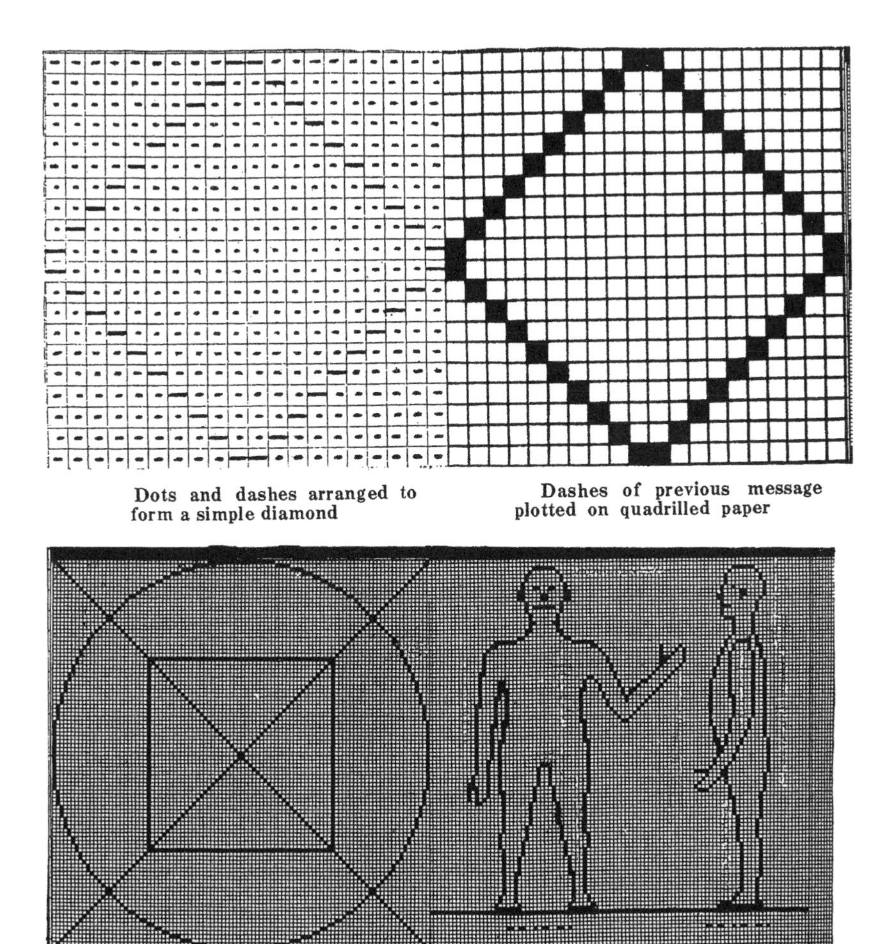

FIG. 8.2. In 1920 the Nieman brothers proposed sending a coded wireless message to Mars. They used dots and dashes to create images. (H. W. and C. Wells Nieman, "What Shall We Say to Mars?" *Scientific American*, March 20, 1920. Copyright 1920 by Scientific American, Inc. All rights reserved.)

to indicate the end of a block of signals. Martians were expected to convert the flashes into black and white squares. Different sequences of dots and dashes yielded different visual messages.

The Nieman technique of communication enabled senders to transmit simple mathematical figures or more complex pictures. The authors illustrated their article with an arrangement of squares depicting a human figure. The figure is shown in frontal and side views. In the frontal view, the figure's left hand is raised, perhaps to show it bends at the elbow.

Interstellar Visits

Not everyone was content to leave the search for extraterrestrial intelligence in the hands of radio astronomers waiting for incoming alien signals. Scientists critical of an exclusive reliance upon radio astronomy proposed more dynamic ways of learning about extraterrestrial civilizations.

Travel through interstellar space was an alternative but never a very popular one among scientists. As early as 1959, physicist John R. Pierce of the Bell Telephone Laboratories considered the relativistic implications of interstellar space travel. He concluded it was doubtful that a spaceship could carry enough fuel to accelerate it to velocities close to the speed of light. Nobel Laureate Edward Purcell discussed the physics of interstellar travel in 1961 and dismissed trips beyond the solar system as not feasible. "All this stuff about traveling around the universe in space suits," he wrote,"belongs back where it came from, on the cereal box."[12] Purcell proposed two-way radio communication with the inhabitants of other worlds. He argued that it was simpler and cheaper to exchange ideas than objects with our neighbors in space.

The German radio astronomer Sebastian von Hoerner, who worked at Green Bank Observatory with Frank Drake, discussed "manned" interstellar travel in 1962 and drew pessimistic conclusions. In the future, he said, space travel would be limited to our planetary system, and extraterrestrials, facing similar problems, would travel through space in their planetary systems. The problem was time dilation, discussed by Einstein in his special theory of relativity. For example, space travelers moving near the speed of light would travel for 27.3 years according to their reckoning while 1,550 years elapsed on Earth. Under the circumstances, von Hoerner advised that communication with extraterrestrials by radio signals was the preferred method.

In 1963, when most scientists doubted the possibility of interstellar space travel, Carl Sagan embraced the notion enthusiastically. He advocated direct physical contact among galactic communities by means of relativistic interstellar flight. Sagan admitted the drawbacks of time dilation but believed they could be overcome by slowing down the metabolic rate of space crew members. Or, since the inhabitants of the home planet belonged to very long-lived civilizations, successive generations could maintain records of their space ventures.

These records would contain the departure time, destination, and anticipated time of arrival of spacecraft that began their voyages thousands of years earlier. Spacefaring societies, in communication with one another, might even share information about exploring spacecraft to eliminate unnecessary duplication. The situation, Sagan said, resembled post-Renaissance seafaring nations and their colonies before the coming of fast clipper or steam ships. Sagan concluded that "other civilizations, aeons more advanced than ours,"[13] were already traveling between the stars.

Sagan recalculated Drake's equation to determine the number of advanced technological civilizations in the Milky Way Galaxy likely to engage in space travel. He based his calculations on two assumptions. First, that life exists elsewhere in the Galaxy. Second, that levels of intelligence and manipulative ability are of great adaptive value in an organism's evolution.

The lifetimes of extraterrestrial civilizations remained a troublesome problem. The possibility that a nuclear war between the Soviets and the West could destroy all life on Earth caused Sagan to pause in his calculations. He then postulated the lifetimes of technological civilization to be in a range of less than one hundred to more than 100 million years.

Sagan finally determined that technological civilizations in the Galaxy numbered about one million. Therefore, one out of a thousand stars in the heavens had a civilized planet in its vicinity. The closest such civilization would be within several hundred light years of the Earth. A civilization located at this distance was within reach of Sagan's relativistic interstellar spaceships.

Sagan used the results of his solution to Drake's equation to determine the frequency of contact among civilizations within the Galaxy. Sagan's new set of calculations was based on his belief that when a civilization found interstellar spaceflight feasible, it immediately developed the necessary technology no matter how difficult or expensive the undertaking. The scientific advantages gained by a society's contact with other space communities justified the investment in interstellar travel.

Sagan concluded that each star in the Galaxy received a random visit from another galactic civilization at least once every one hundred thousand years. At this visit, space explorers observed which of the star's planets were likely to develop intelligent life. Once a technological civilization appeared in a planetary system, then the frequency of visits increased to one every several thousand years.

This line of reasoning brought Sagan back to the Earth and to several critical questions. If advancing civilizations are monitored periodically, have extraterrestrials visited the Earth in historical times? If so, do historical records show evidence of their visits?

Sagan footnoted his mention of direct contact between extraterrestrials and humans with a reference to Enrico Fermi's cryptic remark, "Where are they?" When Fermi asked this question in 1950, he was referring to aliens which, if they existed, should have been on Earth at that moment. Sagan used Fermi's question to indicate that the possibility of human and alien interaction had been "seriously raised before."[14]

Fermi was a more subtle and careful thinker than Sagan, and it is not clear precisely what he had in mind when he asked, "Where are they?" Most commentators assume that the Italian physicist asked the question ironically. That is why they refer to it as Fermi's paradox. Fermi's paradoxical inquiry is an ironic statement about the gap between the claim that the universe is teeming with intelligent alien life and our failure to find any clues of extraterrestrials on Earth.

Sagan raised the issue of encounters between humans and extraterrestrials in a 1963 article he published in the journal *Planetary and Space Science,* a scientific periodical refereed by his peers. The National Aeronautics and Space Administration funded Sagan's research for this article with a grant. Hence, this is a scientific paper, not a sample of Sagan's science fiction or one of his speculative excursions into popular science.

In this article, Sagan warns that there are no reliable reports of alien contacts with humans during the past few centuries. However, contact may have occurred earlier and the evidence modified and distorted by legend and metaphysical or theological speculation. Despite the difficulty of proving such contact, Sagan refers to some ancient documents that he feels deserve notice.

The texts he names preserve the Babylonian account of the origins of Sumerian civilization in the fourth millennium B.C. According to this account, an event occurred on the shores of the Persian Gulf, near the site of the ancient Sumerian city of Eridu. A strange fish-like creature suddenly appeared to the inhabitants of the region. The creature taught these uncivilized peoples the fundamentals of the arts and sciences, including mathematics. It also instructed them in how to build houses and temples, compile tables of laws, and practice agriculture. Sagan later expanded on this story in his 1966 book *Intelligent Life in the Universe.*

Sagan admitted that he had no hard evidence to prove that the ancient Sumerians owed their newly acquired civilized ways to fish-like visitors from outer space. He also noted that representatives of galactic civilizations might have visited the Earth as many as 10,000 times during the geological era of Earth's history when no human witnesses were present. In that case, the visiting space travelers may have discarded artifacts, which have since vanished or remain undiscovered on Earth, or erected an automatic base in our solar system.

During his discussion of extraterrestrial contact, Sagan was careful not to commit himself to the unequivocal truth of the story he told. However, Sagan's scientific credentials, the authority given his article by NASA funding and its publication in a scientific journal, and the author's elusive language helped to legitimize and establish his interpretation of the ancient texts. In a different setting, knowledgeable readers would simply dismiss his story of Sumerians and visitors from outer space as nonsense.

No specialist in Sumerian studies ever claimed that Sumerian civilization began when extraterrestrial visitors passed on their superior knowledge to the primitive inhabitants of Lower Mesopotamia. Furthermore, the creation stories of many cultures tell of superior creatures, usually divinities, bringing the gifts of fire, learning, agriculture, or technology to humanity. Did visitors from outer space routinely confer the blessings of civilization upon so many different peoples? Or are we merely uncovering myths held in common around the world?

In 1960 astronomer Thomas Gold argued that space travelers had visited the Earth billions of years earlier than Sagan had supposed. He presented these views in an article entitled "Cosmic Garbage." Gold proposed that these early visitors accidentally introduced life to Earth in the garbage they left behind when they moved on to other parts of the universe. He noted that when space travel becomes common, humans will likewise spread life by carrying microbes to lifeless planets.

Sagan and Gold spoke with the authority of science. Gold, who was at one time associated with the Harvard College Observatory, was chair of the department of astronomy at Cornell University and director of its Center for Radiophysics and Space Research. He convinced Cornell University to hire Carl Sagan and make him director of a new Laboratory for Planetary Science.

Astroengineering and Supercivilizations

Freeman Dyson was among the few scientists who defended the feasibility of interstellar travel. In 1964 Dyson, a distinguished theoretical physicist and professor at the Princeton Institute for Advanced Studies, envisioned voyages by nuclear propelled spacecraft moving at slow speeds and lasting thousands of years. The crews of such spacecraft would be frozen, placed in cold storage, and revived when needed. Travel that extended over a millennium might not appeal to humans but, as Dyson said, we have no right to impose our taste on others. Since we already have nuclear power sources, interstellar travel is more a biological than a physical problem.

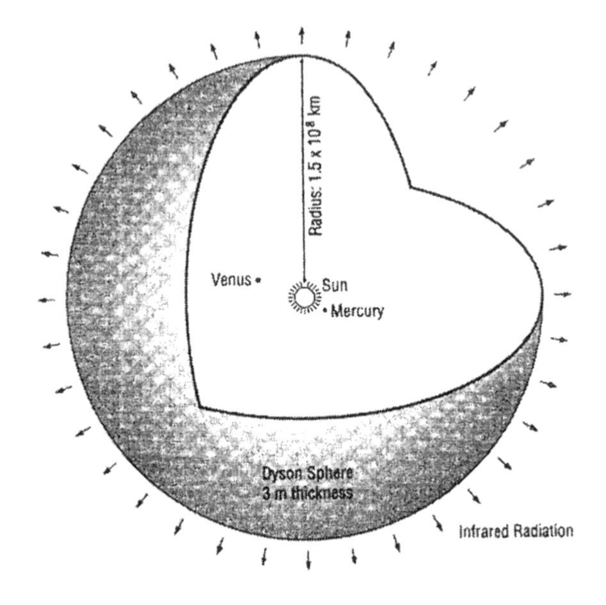

FIG. 8.3. A schematic drawing of a Dyson sphere fashioned from the mass of Jupiter. Its radius is one astronomical unit, the mean distance from the Earth to the Sun. (G. A. Lemarchand, "Detectability of Extraterrestrial Activities." *SETIQuest*, vol. I, no. 1, 1994. Copyright 1994 Helmers Publishing, Inc.)

Despite his interest in space travel, Dyson is better known for postulating the existence of Dyson spheres in regions far from our solar system. A Dyson sphere is a huge artificial sphere built by an extraterrestrial civilization. This gigantic astroengineering project encloses the sun and planet of its builders and captures all of the sun's energy for their exclusive use (Fig. 8.3).

Dyson assumed that it was "overwhelmingly probable" that any extraterrestrial civilization we contacted would be millions of years old. The technological level of such a civilization, he imagined, far surpassed "ours by many orders of magnitude."[15] Given these assumptions, Dyson concluded that most alien civilizations had "expanded to the limits set by Malthusian principles." Hence, they suffered from overpopulation and a shortage of material resources and energy, problems that plagued terrestrial civilizations. Dyson went on to explain how extraterrestrials handled these serious threats.

Planetary systems consist of a central star, or sun, and orbiting planets. These essential elements are available to advanced civilizations. It is reasonable to expect, Dyson wrote, that by the time any technological civilization is a few thousand years old, "Malthusian pressures" will force its citizens to make the most effective use of available resources. By capturing all the energy of its central star, and appropriating other planets for building materials, the hard-pressed extraterrestrials will construct "an artificial biosphere."[16] This gigantic structure encloses the inhabited planet and its parent star. The proposed biosphere is not a solid shell. It is a "loose collection or swarm of objects"[17] independently orbiting the star.

A Dyson sphere, built from matter taken from other planets in its solar system, captures its star's energy for use by the inhabitants of the biosphere. Dyson calculated that engineers using the matter in a Jupiter-sized planet could build a sphere with a radius of over 150 million kilometers.

The inner surface of a Dyson sphere reflects solar radiation toward its center. The conserved radiation provides ample energy for the fast-growing population living within. Although the sphere is huge, it is not visible to distant observers because it is not bright. Waste heat, however, continually escapes from the outer surface of the structure. Dyson determined that an alien-built sphere could be detected by the infrared radiation coming from its surface. Hence, he asked astronomers using sensitive detectors to scan the skies for artificial objects emitting infrared radiation within certain limits.

In the 1960 paper that Dyson prepared for the professional audience of the journal *Science*, he did not reveal the source of his idea of artificial spheres. However, he was more open about its origins in his memoirs published two decades later. There he acknowledged that he got the idea from an old science fiction novel, *Star Maker* (1937) by Olaf Stapledon. In *Star Maker*, Dyson read that in the future "a gauze of light traps" focused "escaping solar energy for intelligent use."[18] Whole galaxies grew dim as Stapledon's aliens built light traps around stars.

In a 1984 interview, Dyson discussed the nature of extraterrestrial intelligence. He mentioned the Cocconi and Morrison paper as the source of a narrow approach to the subject. After the paper's publication, he complained, scientists tended to think that aliens were essentially like us technologically and that they communicated using radio signals.

Dyson cited his own work as an example of a new departure. He claimed that he did not base his artificial biospheres upon any detailed assumptions about the nature of extraterrestrial life. Instead, he considered the consequences of high technology that follow from fundamental physical principles.

Dyson made explicit assumptions about alien life although he claimed he had not. In his *Science* article, he assumed that technological civilizations existed elsewhere in the universe, that those civilizations were similar to Western civilization except they were much older and more advanced, and that extraterrestrial civilizations faced the same pressures from population expansion that operated on Earth.

The assumptions Dyson made about the existence of intelligent life in the universe, the nature of civilization, the role of technology in intelligent life, and the part played by Malthusian principles in the growth of civilizations are essential to his analysis. Calculating the size of a sphere that could be built from the matter in a large planet is easy when compared to determining the causes of the origin, growth, and collapse of civilizations, and the role played by technology in the process.

Dyson, like many other physical scientists, seems unaware of the complexity of the assumptions he brings to the topic. His ideas about extraterrestrial civilizations raise controversial historical issues that cannot be resolved by an appeal to the laws of physics. For instance, the relevance of Malthusian principles to human populations is a contentious subject currently under discussion in the social sciences. Its extension to alleged extraterrestrial populations is highly problematical.

Dyson revisited the idea of artificial biospheres in a 1966 essay he wrote, "The Search for Extraterrestrial Technology." This essay appeared in *Perspectives in Modern Physics*, a volume honoring the distinguished physicist Hans C. Bethe on his sixtieth birthday. Here Dyson restated his claim that extraterrestrial civilizations would indulge in big engineering projects, such as constructing artificial biospheres, rather than concentrate upon sending radio signals to the rest of the universe. However, he reluctantly concluded that expanding technology never "really got loose in our galaxy." If it had, we would find starlight "carefully dammed and regulated" and stars "grouped and organized."[19] Although he was skeptical about the existence of large-scale extraterrestrial technology projects, Dyson believed that the search for them should continue.

Dyson's grand program of infrared searches for proof of his supercivilizations found few supporters among astronomers. However, it inspired Soviet astrophysicist Nicolai S. Kardashev (1932–) to propose a hierarchical classification of the technological civilizations inhabiting the universe.

Kardashev was a student of Shklovskii and a pioneer in the Soviet Union's search for extraterrestrial intelligence. As a young man, he was deeply influenced by Flammarion and Schiaparelli's writings on Martian life. In 1964 Kardashev published a paper in *Soviet Astronomy* on "The Transmission of Information by Extraterrestrial Civilizations." Here he considered the amount of energy required to send coded radio signals at great distances across the universe.

Kardashev's calculation of the energy needed to transmit vast quantities of information from more highly developed civilizations to ones less developed led him to postulate the existence of three types of technological civilizations. The amount of energy it controlled determined the type of each civilization.

A Type I Kardashev civilization is similar to the modern technological societies found on Earth. It draws upon the energy falling upon a planet from its sun. Kardashev estimated the Earth's energy consumption at about 4×10^{19} ergs per second. The Earth has not quite reached Type I status because its inhabitants are unable to capture all of the radiant energy streaming down upon it. For this reason, Carl Sagan said that the Earth was more accurately called a Type .7 civilization.

Kardashev believed that given the number of stars in our Galaxy, and the number of galaxies in the universe, there must exist civilizations much older than any terrestrial civilization. Reflecting upon the Earth's progress in producing more energy over time, Kardashev argued that civilizations billions of years old must possess the ability to control enormous amounts of energy.

Kardashev's ancient and technologically superior civilizations fall into two classes. A Type II supercivilization is able to capture all the radiant energy emitted by its sun for its own technological purposes. In essence this civilization has constructed a Dyson sphere to ensure that almost no solar energy escapes into space. Energy consumption in this case is roughly 4×10^{33} ergs per second.

There is a large gap between a Type II and Type III civilization because the latter has gained control over the total energy output of its galaxy. This gigantic technological achievement calls for the utilization of the power of billions of stars. Therefore, its energy consumption approaches 4×10^{44} ergs per second.

Kardashev believed that terrestrial radio astronomers had a very slight chance of contacting a Type I civilization. It was possible, however, to detect and receive information from Types II and III supercivilizations. He maintained that the two higher civilizations produced immense quantities of information that they broadcast continuously. Unlike Dyson, who advocated searches for radiation leaking from biospheres, Kardashev believed his supercivilizations were using the energy available to them to send coded signals throughout their galaxy and the universe. However, like Dyson, Kardashev was caught in the dilemma posed by Fermi's paradox.

Where are the billion-year-old supercivilizations who are attempting to contact intelligent life elsewhere in the universe? Is there any evidence of very powerful artificial radio sources in the universe? Kardashev claimed that such evidence might exist. Astronomers at California Institute of Technology had recently detected two sources of radio-frequency emission; they catalogued them as CTA-21 and CTA-102. These two met Kardashev's specifications for an artificial radio source.

During a 1981 interview, Kardashev repeated his belief in the three types of technological civilizations. He also acknowledged that his mentor Shklovskii no longer believed in the existence of intelligent extraterrestrial life. Finally, he conceded that CTA-102, which he and other Soviet astronomers once publicly hailed as proof of the existence of a supercivilization in outer space, was a distant quasar with a very large red shift.

Despite its lack of empirical verification, Kardashev's scheme for classifying extraterrestrial civilizations joined Frank Drake's equation as a way of thinking about intelligent extraterrestrial life. Carl Sagan introduced Kardashev's ideas to American scientists in an article on the detection of advanced galactic civilizations in 1973. Sagan urged radio astronomers to search for Type II or III supercivilizations among the nearer galaxies rather than Type I or younger civilizations among the nearer stars.

• • •

The acceptance of the idea of an expanding universe, the introduction of the tools and techniques of radio astronomy, and research into the chemical basis of life helped to make interstellar space the new realm for speculation about advanced extraterrestrial civilizations. Drake's equation and Kardashev's formulation of supercivilizations took advantage of the new vistas open for speculation. The parochialism that limited advanced life to our solar system was gone. Now scientists discussed the pros and cons of interstellar travel, earlier visits by extraterrestrial beings to Earth, and the possibility of identifying the astroengineering projects of extraterrestrial civilizations.

The tendency to think about advanced life forms inhabiting outer space was enhanced by the late twentieth-century discovery of extrasolar planets, bodies that orbited distant stars. This discovery provided observational evidence for the second factor in Drake's equation. The Earth was not unique. Its patterns of life and culture might be duplicated on worlds located many light years in the distance.

CHAPTER NINE

The Trajectory

CETI to SETI to HRMS

> . . . when I address the floor tomorrow . . . we will not be talking about SETI . . . we will be talking about HRMS, which is the new name by which this program continues to have life. And it will be my intention, once again, to offer an amendment which specifically deletes the funding for this program.
>
> —Senator Richard Bryan, *Congressional Record*, September 20, 1993

The Birth of CETI

The search for extraterrestrial intelligence had Russian as well as American origins. The early successes of the USSR space program, an interest in the subject by Soviet astronomers, and the Marxist materialistic philosophy that stressed the physical basis of life inspired Soviet scientists to hunt for advanced alien civilizations. Soviet-era scientists also drew upon the engineering and philosophical work of the Russian rocket pioneer Konstantin E. Tsiolkovsky (1875–1935). Apart from his contributions to space science and technology,

Tsiolkovsky also speculated about advanced intelligent life on other planets and attempted to account for our failure to find any evidence of it. Twenty-seven years after Tsiolkovsky's death, the Soviet astrophysicist I. S. Shklovskii published an influential Russian language book on the search for extraterrestrial intelligence. With the help of Carl Sagan, an expanded and revised English translation of the volume appeared in 1966. The title of the new edition was *Intelligent Life in the Universe*. This work introduced the subject to American scientists and was popular with the wider public.

The Soviet Union sponsored a conference on extraterrestrial civilizations and interstellar communication in 1964 at the Byurakan Astrophysical Observatory, Armenia. This meeting, limited to Soviet scientists, featured Shklovskii and Kardashev as speakers. The former was interested in the expansive nature of supercivilizations. If a supercivilization was able to fabricate a Dyson sphere, he said, it could subdue an entire galaxy in a few tens of million years. Shklovskii admitted that genetic failure, thermonuclear war, or some other catastrophe might halt the galactic expansion of a supercivilization.

It was possible that a supercivilization might not care to communicate with inferior societies on Earth. Nevertheless, in 1964 Shklovskii was confident that advanced extraterrestrial civilizations would attempt to contact others in the universe using radio waves. Shklovskii's portrayal of advanced life, and the possibility of communicating with it, was accepted by his fellow conferees.

Kardashev claimed that by using existing radio telescope equipment we had a better chance of contacting Types II and III civilizations than those at a lower level. He assumed that an Earthlike civilization (Type I) would develop within a few billion years of the beginning of life on the planet. A Type I civilization might advance to a Type II within a few more thousand years. Type III civilizations, however, were likely to appear tens of million years later. The key feature of supercivilizations was the vast energy resources at their command. Kardashev supposed that supercivilizations would use these resources to transmit large quantities of information across interstellar distances.

A. V. Gladkii, a Soviet mathematician, gave the final talk at Byurakan. He noted that a cosmic language already existed for communicating with supercivilizations. In 1960 Dutch mathematician Hans Freudenthal published LINCOS, an elaborate language based upon mathematical symbols and designed for cosmic communication. Gladkii and Freudenthal assumed that mathematics would be understood by all intelligent creatures in the universe.

The Byurakan conference ended with a call for the establishment of interstellar communications. This meant sending radio signals from Earth and searching for signs of incoming messages from alien civilizations. Scientists attending the conference agreed that signals transmitted by extraterrestrial civilizations would contain information beneficial to the natural sciences, philosophy, and everyday life on Earth.

In early September 1971, the Byurakan Observatory was the site of a second meeting on alien communication. This time it hosted the first international conference on communication with extraterrestrial intelligence (CETI). Carl Sagan and Nikolai Kardashev proposed the largely Soviet-American CETI conference. The prestigious American National Academy of Sciences and the Academy of Sciences of the Soviet Union cosponsored the affair. As a gesture of goodwill, the Soviets gave souvenir pins to the attendees inscribed with the letters CETI. This gift helped to establish CETI as an internationally recognized acronym.

The second Byurakan conference included astronomers, physicists, biologists, chemists, social scientists, philosophers, and historians. Kardashev originally intended to restrict participation to astronomers and physicists. He hoped to exclude philosophers and other nonscientists, whom he derisively called "windbags."

The conference set aside several days to discuss the individual elements that made up Drake's equation. In the open discussion that followed, historian William H. McNeil of the University of Chicago noted that the number of factors in Drake's equation was arbitrary. Suppose we add new factors to the equation, he asked, factors that may have a value less than one? In that case, the number of communicating civilizations in the Galaxy (N) would shrink.

Sagan challenged McNeil to add an extra hundred or thousand factors to the equation, and if each additional factor had a value of one, then N would remain large. McNeil countered with his original argument that not all factors necessarily had a value of one. After all, conference members were working to determine the *probability* of each of Drake's factors, a probability that might be less than one.

Sagan replied that any increase in factors would lead to the same result, a technological civilization roughly equivalent to ours. All evolutionary paths converged on the same goal. Several scientists rallied to Sagan's defense without actually responding to McNeil's criticism.

In essence McNeil asked the assembled scientists to rethink Drake's equation by justifying the choice and number of factors included. It was possible to draw up a different, and longer, list of factors affecting the number of advanced Galactic civilizations. McNeil, the historian, saw complexity, contingency, and accident where the scientists saw deterministic paths leading to technological civilizations in the Galaxy.

The conference inevitably turned to the determination of the number (N) of advanced civilizations in the Galaxy. A cautious Sagan revised the optimistic estimates proposed at the Green Bank conference in 1961. He now suggested that the life span of an advanced civilization (L) was 10 million years and that N was roughly 1 million, or one civilization for every 100,000 stars. Sagan admitted that his evaluation of L might be too high. He and others listed the many problems faced by technological civilizations. They wondered if these societies could overcome the combined threat of nuclear warfare, environmental pollution, and overpopulation to survive for 10 million years.

Freeman Dyson opened the next session (astroengineering activity) by declaring, "to hell with philosophy. I came here to learn about observations and instruments."[1] Curiously, Dyson discussed neither of these topics. Instead, he hypothesized the existence of Galactic comets and claimed they might be habitats for intelligent life.

In the lively discussion inspired by Dyson's declarations, physicist Charles Townes said he understood that comets in the Galaxy could provide raw materials for civilizations but not why they were suitable sites for advanced life. Dyson answered that the sheer number of Galactic comets provides the largest living space we know and "that you can get away from your government." It is unclear which governments Dyson had in mind—perhaps a society in a Dyson sphere or one of Kardashev's supercivilizations.

The participants entertained ever more wild ideas thereafter. If the tidal stresses near the entrance of black holes were dangerous, said Sagan, they would create navigational hazards for space travelers. Interstellar civilizations might surround these danger spots with navigational buoys to facilitate safe travel. Sagan thought that the buoys might be detectable by terrestrial observers.

The conference then debated the consequences of contact with extraterrestrials. Philip Morrison maintained that experts could not decode and interpret an alien signal quickly. Instead, interdisciplinary teams working together for many years would determine the meaning of an extraterrestrial communiqué.

Morrison compared the task of interpretation to the work of scholars who over the centuries deciphered the complex message embedded in the history and culture of ancient Greek civilization. He predicted that an extraterrestrial message would appear as "a discipline rather than a headline or an oracle."[2]

William McNeil, who had not spoken since his comments on Drake's equation, rejoined the discussion. McNeil regretted Morrison's uncritical approach to the question of extraterrestrial communication. He also doubted that humans had the ability to decode alien signals. Our intelligence, he argued, is a prisoner of the language we have devised. McNeil suspected that the language of intelligent aliens would have few, if any, points of contact with human language. Therefore, he doubted if any useful information could cross the divide separating terrestrial and extraterrestrial civilizations.

Francis Crick responded that the universal language of mathematics made it a natural link between aliens and humans. You are not justified, McNeil countered, to claim that our mathematics is their mathematics. Crick retorted that the assembled mathematicians would agree with him, not McNeil.

McNeil then made a general comment about the unrestrained speculation that swirled around him. "I feel I detect," he said, "what might be called a pseudo or scientific religion" in the making. McNeil did not criticize his colleagues for their deeply held beliefs. Faith, hope, and trust, he said, are fundamental to humanity, and it would be wrong to dismiss those who cling to them. Nevertheless, he concluded, "I remain . . . an agnostic, not only in traditional religion but also in this new one."[3] When McNeil returned from the Byurakan conference, he summarized his experience there in an article published in the University of Chicago alumni magazine. The piece was entitled "Journey from Common Sense."

Bernard Oliver, vice-president for research and development at Hewlett-Packard, told the Byurakan audience that McNeil's critical response reflected the opinion of an educated person who does not know much about science. Oliver went on to remind the historian that intelligent extraterrestrials very likely have eyes and are thus able to decode pictorial messages sent from Earth. Oliver said nothing about the neurological, mental, intellectual, and cultural factors involved in decoding visual data.

McNeil touched a sensitive spot with his comments on the birth of a new scientific religion and the group's mathematical bias. Neither Crick nor his mathematical cohorts confronted the thorny problems raised by the idea of mathematics as a language understood everywhere in the universe. The supposed universality of mathematics is an *assumption* about the nature of mathematics not shared by all professional mathematicians nor all philosophers of mathematics. The topic remains an unresolved philosophical dispute. It is true that our mathematics is applicable to phenomena in distant and vast regions of the universe. However, that is not proof that intelligent creatures could not have developed different systems of mathematics, or other forms of knowledge, to deal with known or unknown phenomena.

Oliver attributed McNeil's failure to endorse the belief in extraterrestrial intelligence to his lack of scientific training. He also hinted mysteriously that such talk had political ramifications. In responding to Oliver's remarks, a critical commentator might ask: Was it Sagan's scientific background that led him to expect to find navigational buoys around black holes and to estimate the lifetimes of nonexistent civilizations? Was it Kardashev and Shklovskii's advanced study of astronomy that inspired them to discuss and classify supercivilizations that neither they nor anyone else had ever observed? Finally, was it Oliver's study of the physical sciences that permitted him to grant imaginary extraterrestrial beings the ability to decipher terrestrial languages and codes? In retrospect, McNeil's doubts were more perceptive than the speculative accounts generated by his critics.

The conference closed with some final remarks by Morrison. He said that in the long run posterity would remember the 1971 conference as the event that made the search for extraterrestrial intelligence scientifically respectable. Two years later, Sagan recalled Byurakan as a turning point in the search for extraterrestrial intelligence.

From CETI to SETI

While scientists debated CETI in the USSR, the leaders of America's space program were modestly interested in finding evidence of extraterrestrial intelligence. Initially, NASA focused upon planetary exploration and life within the borders of the solar system. By 1970 the space agency sponsored small interstellar communications study groups featuring people like Sagan, Drake, Oliver, and Morrison. The success of this venture led to a jointly sponsored NASA-Stanford University engineering faculty program. The result was Project Cyclops.

Project Cyclops called for an "orchard" of several thousand large dish antennae symmetrically arrayed over an area ten miles in diameter (Fig. 9.1). Engineers designed Cyclops antennae to explore a portion of the radio frequency band in which noise coming from natural celestial sources was at a minimum. This band lies between the twenty-one-centimeter line of hydrogen (H) and the eighteen-centimeter line of the hydroxyl radical (OH). Since H plus OH yields water (H_2O), Oliver called the chosen region the "waterhole." The cosmic waterhole was, he said, an appropriate place for all manner of life to gather together. In outer space, as on Earth, the waterhole was a common meeting ground for a variety of creatures.

Cyclops was ambitious and expensive, with a projected cost of 6 to 10 billion dollars. The space agency never built the proposed array of antennae. However, the project report included a crucial recommendation. It urged NASA to make the search for extraterrestrial intelligence an ongoing part of the space program, with a budget and ample funding. NASA officials accepted this recommendation and began limited, but regular, financial support for research on interstellar communication.

In 1975, NASA sponsored a series of science workshops on the search for extraterrestrial intelligence. Philip Morrison chaired the series. These workshops

FIG. 9.1. Artist's conception of an array of Cyclops system antennae. (John Billingham and Bernard M. Oliver, *Project Cyclops*. NASA/Ames Research Center, 1972.)

gained support from the wider scientific community and laid the foundation for future NASA-sponsored research. NASA did not intend to dispatch any new signals into space to add to those already transmitted by television broadcasters and radar installations over the years. Instead of active communication with other worlds, NASA opted for passive searches for evidence of incoming signals from space.

Sending messages into space was a controversial activity, as Sagan and Drake discovered in 1974 when they used the Arecibo radio telescope to deliver a message to a star cluster. Critics expressed concern that hostile aliens might intercept the message, determine its source, and possibly invade the Earth. As Drake noted, by the mid-1970s it seemed arrogant to stress communication (CETI) when there was yet no one to whom one might send a radio message. Search, not communication, was the modest goal of interested radio astronomers. For these reasons, NASA chose SETI over CETI to designate its new interest in extraterrestrial intelligence.

Where Are They?

The early years of NASA's interest in SETI coincided with the failure of the Viking Landers in 1976 to find any hint of life on Mars. In the aftermath of NASA's inability to discover Martian life, the University of Maryland sponsored a symposium entitled "Where Are They?" in 1979. Astronomers Michael H. Hart and Benjamin Zuckerman directed the event. They decided to revisit Enrico Fermi's paradoxical question about the whereabouts of aliens.

The published proceedings of the Maryland symposium opened with a critical piece Hart had written earlier (1975). The main point of Hart's essay, what he called Fact A, is that there are no intelligent beings from outer space currently on Earth. Hart was among the first to make colonization of space by aliens a central issue in the SETI debate. He, and others at the symposium, argued that interstellar travel by technologically advanced civilizations would lead them to establish colonies on the habitable planets they visited. The space colonization analogy drew upon the history of early European explorers who traveled to the New World and established settlements there.

Hart methodically dismissed arguments offered to explain the failure of aliens to visit and settle the Earth. He ruled out physical explanations by showing that there are no insuperable obstacles to travel through outer space. Space flight at a velocity of one-tenth the speed of light is fast enough to bring interstellar craft into our vicinity.

Hart next dealt with sociological explanations for the absence of aliens. These included claims that space exploration holds no interest for advanced civilizations; technologically advanced civilizations have destroyed themselves in nuclear wars; or aliens have set aside the Earth as a wildlife preserve or zoo. The zoo hypothesis was not new. It had a long tradition in science fiction and was discussed by Tsiolkovsky early in the twentieth century. In the 1970s, several scientists proposed that terrestrial life is preserved in a natural state and scrutinized periodically by creatures superior to us.

Hart maintained that the zoo hypothesis and other sociological explanations are flawed. The assumptions behind these explanations may hold true for *some* extraterrestrial civilizations but not for every one of them. For instance, a few extraterrestrial civilizations might destroy themselves in a nuclear holocaust, but that is not the fate of others. Or, a few extraterrestrial civilizations might play the role of zookeepers of the Earth, but not all take on that responsibility. If life is abundant in the Galaxy, then the possibilities are endless, and explanations of this sort cannot account for missing space colonies.

Hart went on to draw several conclusions from his analysis. First, we are the earliest creatures to become civilized in the Galaxy. Second, electronic searches for extraterrestrial intelligence are a waste of time and energy. Third, humans eventually will colonize the empty Galaxy.

The various participants at the Maryland conference responded differently to the implications of Fermi's paradox. Nuclear physicist Eric Jones ran a computer simulation of the expansive nature of Galactic civilizations and determined that extraterrestrials should be on Earth now. Clearly, that was not the case. Therefore, other speakers tried to explain this situation. No one endorsed the idea of occasional visits by UFOs.

Astronomer Michael Papagiannis suggested that interstellar beings had established space colonies in the asteroid belt that orbited the Sun between Mars and Jupiter. The aliens hid their colonies among the belt's many natural objects. Papagiannis asked astronomers to search the asteroid belt for irregularities that might disclose the presence of artificial structures.

Freeman Dyson spoke on interstellar propulsion systems. The braking systems used by interstellar spacecraft held a special interest for Dyson. He calculated that a space vehicle braking from a very high velocity would leave a long trail of plasma (hot ionized gases) in the sky. Astronomers using radio telescopes might detect the plasma skid marks left by nearby spacecraft.

Hart ended the symposium with the assertion that in an infinite universe there are undoubtedly an infinite number of inhabited planets. However, the chance that life exists in a given galaxy is very small. Therefore, intelligent beings are unlikely to be able to contact another advanced race in their galaxy.

Hart wondered why, despite evidence to the contrary, so many people accept the idea that there are a large number of advanced civilizations in the Galaxy. He decided that it was simply a matter of wishful thinking. Humans preferred to imagine a Galaxy filled with strange life forms to one in which they are alone.

While scientists at the Maryland symposium reviewed the Fermi paradox, Frank Tipler, a professor of physics and mathematics at Tulane University, prepared an assault on the SETI hypothesis and its reliance upon the detection of alien radio signals. He bluntly announced that intelligent extraterrestrial beings do not exist. Tipler's 1981 article appeared in the *Quarterly Journal of the Royal Astronomical Society* after it was rejected by several other scientific journals. According to some accounts, Carl Sagan may have used his influence to block its earlier publication.

Tipler noted that if intelligent extraterrestrial beings exist, their spaceships have already penetrated the solar system. Advanced civilizations capable of beaming radio signals to Earth could easily master interstellar travel. If these civilizations were truly superior to ours, they would have sophisticated computer technology and the ability to build and dispatch self-reproducing exploration probes. In 1966, mathematician John von Neumann established the theoretical basis for such machines. He called them self-reproducing automata.

A von Neumann machine can make a copy of itself using available materials, such as an asteroid or cosmic debris. The copy travels to another location in the universe, where the process begins again. Distant aliens could launch von Neumann space probes to explore the Galaxy. The probes would automatically relay information to their home base as they carried out a preprogrammed exploration strategy. Tipler estimated that von Neumann probes would fill the entire Galaxy within 300 million years.

The absence of space probes in the Earth's vicinity led Tipler to recalculate Drake's equation based on his set of assumptions for the different factors. Given the observational fact that there is no sign of extraterrestrial activity, Tipler determined that the number of civilizations in our Galaxy is one. If we are alone, Tipler asked, why do we try so hard to prove otherwise? His unusual explanation was that the scientists' search for extraterrestrial intelligence had much in common with the popular belief in UFOs. In both cases, people believe that humans can save themselves by the miraculous intervention of interstellar beings. To support this allegation, Tipler cited an article Frank Drake published a few years earlier.

The Immortals

Drake's article appeared in a 1976 issue of *Technology Review* celebrating Philip Morrison's sixtieth birthday. He cryptically entitled it "On Hands and Knees in Search of Elysium." Drake conceded that radio telescope searches had yielded no results so far, but he thought this was not extraordinary. Humans must expect to work for the priceless knowledge possessed by their intellectual superiors.

At this point, Drake flatly asserted that any extraterrestrial intelligences we contact are likely to be immortal. Immortals are not rare, Drake declared. They may even be very common in the universe.

Drake's idea of immortality was more material than spiritual. He envisioned the elimination of the aging process and the transfer of memory from older to younger brains or to the brains of clones. Drake imagined that aliens lived in a medical utopia, a society forever free of disease.

Drake believed that in the future humans will acquire physical immortality. Just as nuclear energy and radio telescopes are inevitable in the development of technological civilization, so is immortality inevitable in the biological realm. Contemporary medical researchers, who cured polio and other malignant diseases, were equipped to put humanity on the road to immortality.

Modern medical researchers discredit Drake's utopian idea of a disease-free world. They argue that there are not a fixed number of diseases that can be eliminated one by one over time. Disease is part of the same evolutionary process that produced the human race. We evolved together and continue to do so. Medical researchers might wipe out some diseases, but others soon rise to replace them. A generation that has experienced the worldwide spread of AIDS, the ravages of the Ebola virus, the threat of the human form of Mad Cow Disease (Creutzfeld-Jakob variant), the appearance of SARS, and the possibility of an influenza epidemic appreciates how improbable it is to imagine immortality of the kind proposed by Drake.

To return to Drake's scenario. Immortal aliens have conquered disease, but they are vulnerable to fatal injuries. For that reason, personal safety is of great concern to immortals. They avoid travel on crash-prone aircraft and vigorously oppose wars. Immortals reduce the possibility of war by freely distributing the secret of longevity to creatures living at lower stages of technological development. The need of immortals to send this message results in the transmission of a large number of radio signals.

Immortal beings are obviously not in a hurry. Hence, they develop techniques of signaling that reach all regions of the Galaxy using low power and a very limited frequency band. In the past, terrestrial radio telescope operators searched broader bandwidths and missed messages sent on narrower bands. According to Drake, this is a serious mistake. Immortal civilizations are in the majority, and they send messages frequently. SETI researchers must listen for narrow-band signals transmitted through interstellar space.

Drake closed his essay by imagining an event twenty years in the future. The time is 1996. The place is a listening post in the Mojave Desert filled with rows of antennae scanning the heavens. After a decade of search, a message finally arrives. The message, in binary code, continues for a year before scientists decipher it.

The messengers are immortals who have been alive for a billion years. They are now ready to share their formula for immortality with their technological inferiors on Earth. Should humans accept the formula or reject it? Drake leaves the question unanswered.

When Frank Drake proposed the immortality of extraterrestrials, he was Director of the National Astronomy and Ionosphere Center and a chaired professor of astronomy at Cornell University. His essay appeared in a journal published by the Massachusetts Institute of Technology honoring one of its well-known faculty members. Consequently, it carried considerable scientific weight.

Fifteen years after Drake first insisted that aliens were immortal, he repeated his claim in a book on the scientific search for extraterrestrial intelligence. Drake's ideas on the subject changed little over the intervening years except that he now hypothesized that in order to prevent deadly accidents, vehicles carrying the immortals moved very slowly. The vehicular speed limit was slightly above zero. Drake's second major discussion of immortals ended with him staring at the night skies wondering about the interstellar messages streaming to Earth. The most common one, he decided, was a book-length message instructing humans how to live forever.

Tipler called attention to SETI scientists who were anxious to save humanity through the miraculous intervention of extraterrestrials. Drake's writings confirm Tipler's claim and call attention to the long religious tradition that preceded and nurtured the scientific pursuit of imaginary beings. They also recall Drake's remarks about the importance of religious fundamentalism, at an early age, as the initial motivation for himself and other SETI researchers.

Rallying Cry

Frank Drake to the contrary, the absence of either space probes or colonies convinced many that the basic premises of SETI were suspect. After all, the recently established American space program had already launched its version of space probes. Sagan, Drake, and others helped NASA attach messages to the Pioneer and Voyager spacecraft (1972–1977). If NASA engineers could send spacecraft on interstellar journeys, why were technologically superior extraterrestrials unable to do likewise?

As criticism of SETI mounted, Shklovskii was the first important SETI pioneer to change his mind about extraterrestrial intelligence. He had promoted the Soviet search for intelligent aliens, collaborated with Sagan on a classic SETI book, and taught Kardashev astrophysics at Moscow University. Then, in 1975, Shklovskii announced that terrestrial life was unique in the Galaxy.

Shklovskii decided that the evolution of life on Earth and the subsequent appearance of technological civilizations were the coincidental result of a series of highly improbable events. He admitted that some intelligent species might exist in other galaxies. Perhaps these creatures preferred not to travel beyond their territory. In any case, the result was the same. Humans are alone in the Galaxy.

Drake's equation also came under attack during this time. Some commentators proposed modifications in the equation and urged its retention. Others used the equation to prove that the number of communicating civilizations in the Galaxy was exactly one, human civilization. Still others, like Freeman Dyson, rejected the equation as worthless. Dyson opposed any attempt to derive the probability of alien life from a set of theoretical principles.

At a 1979 meeting of the International Union of Astronomers, Philip Morrison answered the growing number of SETI critics. He singled out a flaw in the space colony argument. Morrison claimed that the expectation of space visits, probes, and colonies rested upon the false application of Malthusian exponential growth to technology.

The followers of Fermi and Hart, Morrison alleged, first assume that technology grows exponentially. Then they ask where are the space probes of expanding technological civilizations. Morrison responded that in the real world technology, as well as many other human endeavors, has limited growth.

Morrison chose Nicolaus Copernicus and the principle of mediocrity over Thomas Malthus and exponential growth. He embraced a modern version of the principle that claimed that human society and technology generally represents what exists elsewhere in the universe. Nevertheless, Morrison believed that "we can look for creatures better than ourselves."[4] These creatures do not control stars or entire galaxies. They operate within bounded domains, just as humans do.

SETI's critics argued that if aliens exist, the Galaxy should be filled with their space probes. This argument assumed the exponential expansion of technology. However, SETI's critics were not the first to make that assumption. The idea originated with the astronomers and physicists who initially proposed to search for extraterrestrial intelligence. In their 1966 book on intelligent life in the universe, Shklovskii and Sagan determined that there were between fifty thousand and one million technologically advanced civilizations in the Galaxy.

But if there are no superior civilizations in space, how can we expect them to contact us with their powerful radio transmitters? A strict interpretation of the principle of mediocrity would result in a universe filled with civilizations like ours, not quite ready for Type I status. Such civilizations are unlikely to communicate with one another. Perhaps for that reason, Morrison appeared to contradict the key assumption of the principle of mediocrity with the ambiguous remark that there were creatures in the Galaxy "better than ourselves," but bounded nevertheless. The notion that the universe contains creatures better than ourselves is one of the oldest assumptions made by those who search for extraterrestrial intelligence.

Morrison closed his remarks with a rallying cry to the astronomical community. It is legitimate to theorize about the number of advanced extraterrestrial civilizations, he said, but that is not enough. At some point, it is necessary to make observations, to listen for meaningful radio signals amid the background noise of the universe.

Observation is a unifying element in astronomy. Astronomers willing to dispute Drake's equation and question the absence of space probes and colonies heeded Morrison's call. Many agreed to end their bickering over competing theories and get on with their main job of studying the heavens.

On the surface, Morrison's resolution of the SETI crisis is appealing. It is a call for harmony over dissension and observation over unrestrained theorizing. However, his plea includes problems often sidelined in discussions about the search for advanced alien civilizations. The first problem arises from questions in the philosophy and methodology of the astronomical sciences. What constitutes evidence in astronomy, and how should astronomers gather it? The second problem is political and financial. Who should pay for the searches conducted by SETI teams? The government? If so, which agencies of the government? The private sector? Which private institutions should fund SETI programs? And finally, how much money should we spend on SETI?

To address the first problem astronomers needed to agree on legitimate observational goals. Earlier, they trained their optical or radio telescopes on a selected celestial object and gathered data to resolve disputes over theory. Astronomers searching for Martian canals used their telescopes, or sent their spacecraft, to map the surface of the planet.

But when the existence of hypothetical extraterrestrial life is under scrutiny, the object to study is not well defined. And, it is not clear that the radio telescope is necessarily the right instrument for the job. The route to alien life through radio telescopes makes a fundamental assumption about the nature of extraterrestrial life. A radio telescope used in a SETI search is not a neutral tool of the trade. With the astronomical object in question, SETI researchers hypothesize an alien technology that accommodates their instruments and then calibrate them to suit their hypothesis. In this fashion, the technological and strategic problems of radio astronomy became prime problems in the search for extraterrestrial intelligence.

A number of astronomers were willing to follow this route and continue their search using radio telescopes. Their willingness to begin work raised the second problem. Who will sponsor expensive surveys of the skies that utilize arrays of radio telescopes integrated with computers to differentiate between incoming signals? Should the money come from private observatories and universities? The National Science Foundation? The National Aeronautics and Space Administration? Although astronomers at Harvard, Ohio State, the University of California-Berkeley, and elsewhere pursued SETI research projects, NASA mounted the most substantial efforts.

SETI at NASA

NASA formally gained Congressional approval for its SETI research program in 1978. A year later, when Congress was ready to allocate $2 million to SETI, Senator William Proxmire awarded NASA his infamous Golden Fleece

Award. The award drew public attention to the wasteful spending of taxpayers' money. Congress reacted by canceling NASA's request for funds to search for extraterrestrial intelligence.

The SETI issue resurfaced in 1982 when NASA asked Congress to renew its budget. NASA officials hoped to persuade Congress that the scientific and political climate had changed in favor of SETI. The Astronomy Survey Committee of the National Academy of Sciences (NAS) recently endorsed research on extraterrestrial intelligence. They recommended spending $20 million over the long term on it. The NAS committee argued that intelligent organisms are as much a part of the universe as stars and galaxies.

William Proxmire spoke in the Senate chamber again to denounce spending public money on a futile search for extraterrestrial intelligence. This time he quoted Frank Tipler's argument that if intelligent beings exist, they should appear in the solar system. There is not a bit of evidence, Proxmire said, that intelligent life inhabits outer space. He suggested Americans stop chasing interstellar conversations and open communication with our neighbors on Earth.

Carl Sagan was the only person in America likely to change Senator Proxmire's mind. In the early 1980s, Sagan was the best-known scientist in the United States, if not the world. His speculations on the evolution of human intelligence, *The Dragons of Eden*, won him the Pulitzer Prize in 1978. Two years later, Sagan conceived and narrated the thirteen-part television series *Cosmos*. *Cosmos* surveyed the universe and its possible life forms, bringing popular support for SETI and widespread praise for its narrator. More people around the world watched *Cosmos* than any other televised science program in history.

Sagan, who knew Proxmire, arranged a Washington meeting with the senator. Sagan patiently explained Drake's equation to Proxmire, emphasizing the significance of the lifetime of technologically advanced civilizations. Since both men detested war, Sagan drew a connection between knowledge gained from alien contact and the survival of the human race. Extraterrestrials who had successfully passed through the stage of nuclear conflict, he said, could teach us how to avoid nuclear wars. Sagan's arguments convinced Senator Proxmire. He agreed to halt his campaign against SETI funding in NASA's budget. Congress voted $1.5 million for SETI projects in 1983.

Carl Sagan did not stop with Proxmire's conversion to his way of thinking. He went on to collect signatures for a petition supporting SETI research on an international scale. Eventually, he gathered the names of sixty-nine distinguished scientists, including seven Nobel laureates.

In the petition, Sagan admitted that some scientists questioned the existence of extraterrestrial beings. These skeptics, he wrote, maintain that there is no evidence of major reworking of the Galaxy and no hint of alien colonies in our midst. By "major reworking" Sagan meant proof of astroengineering projects carried out by intelligent extraterrestrial creatures.

Sagan claimed that arguments directed against the existence of alien intelligence were based on extrapolations that went beyond conditions prevailing on Earth. SETI searches using radio telescopes were a different matter. Sagan argued that he and his colleagues assumed nothing about other civilizations that had not already occurred on Earth.

Sagan's response is correct in a limited sense. His opponents introduced self-reproducing space probes and alien colonization while Sagan assumed aliens communicated with radio technology already available on Earth. However, hidden in Sagan's position was the questionable assumption that intelligent aliens evolve along the same cultural and intellectual track as humans. Extraterrestrial science and technology is in harmony with ours because of parallel evolution in biology, culture, science, and technology.

Sagan's vigorous defense of SETI took a new turn when he decided to write a science fiction novel about the first extraterrestrial message received by humans. He got an unprecedented advance of $2 million from his publisher for the novel *Contact*. The novel was a success when it appeared in 1985. It enlarged Sagan's reputation as America's foremost expositor of space science and advocate of SETI research. *Contact* also made information and hypotheses hitherto limited to the small circle of SETI researchers part of popular culture.

Contact is a novel of ideas. The fundamentals of SETI research are the source for the first set of ideas. The second set features the debate between science and religion. The author made his chief character, Dr. Eleanor Arroway, a SETI researcher skeptical of the claims of religion. Arroway is outspoken in her belief that science offers a more direct route to truth than religion. Commentators have suggested that Sagan based the character of Arroway on himself, his wife Ann Druyan, and NASA radio astronomer Jill C. Tarter.

Sagan's agnostic heroine finally meets the aliens who had contacted Earth earlier. During this encounter, she learns that her extraterrestrial messengers are caretakers for a race of a higher rank of beings who preceded them. The novel ends ambiguously with Arroway's discovery that there is a super-intelligence that existed before the creation of the universe. This message is hidden in the infinitely long series of digits that make up π (pi). The precise nature of this super-intelligence is left unclear. Despite the agnosticism of its main character, the conclusion of *Contact* has strong religious overtones.

Contact was on the *New York Times* bestseller list for six months. The novel's popularity and influence were enhanced when it appeared as a film in 1997 with Jodie Foster playing the role of the dedicated SETI investigator. Sagan used the televised *Cosmos* series and the science fiction novel and film *Contact* to popularize and advance the search for intelligent life beyond our solar system. No other scientist could match his contributions to SETI in this arena.

Sagan was not alone among scientists sympathetic to SETI who understood that NASA needed all the help it could get, both popular and political. In November 1984, Bernard Oliver (Hewlett-Packard) and John Billingham (NASA) joined with Frank Drake and others to found the nonprofit SETI Institute. Its goal was to encourage the search for extraterrestrial life by serving as a subcontractor with low overhead expense for SETI research grants.

The SETI Institute made good use of the money it received. The Institute managed to keep its bureaucracy small and attract talented scientists to work on its projects. It administered millions of dollars of funded research, yet its overhead costs were lower than other space subcontractors.

Despite these positive steps in the public sphere, the SETI program continued to find an uncertain welcome at NASA. There were different opinions about strategies to pursue in searching for intelligent extraterrestrial life and the proper place of the program in NASA's overall structure. In Washington, Congressional funding for SETI was often erratic. Nevertheless, SETI began its modest research and development stage in the early 1980s and finally gained the status of an approved NASA project in 1990. The SETI project now had a budget of $108 million to spend over its ten-year phase of development and operation.

The final stage of NASA SETI's project was scheduled for the year 2000. Many SETI supporters expected extraterrestrial contact well before the coming of the millennium. Frank Drake was one of them. He believed that he and his colleagues were on the brink of a significant discovery. Anticipating this historic event, in 1992 Drake wrote a popular book whose title asked, *Is Anyone Out There*? His answer was positive.

Members of Congress, meanwhile, continued to question the validity of NASA's SETI program. In April 1992, six months before NASA officially opened its historic search for extraterrestrial intelligent life, Representative John J. Duncan moved to strike $13.5 million set aside for SETI in NASA's budget. Duncan complained that the fifty SETI searches made since 1960 yielded no results. He suggested that Congress divert the millions of dollars allotted to SETI to help America's poorest citizens.

Representative George E. Brown answered Duncan's charges by declaring that SETI was a legitimate search for irregularities or anomalies received by radio telescopes scanning the skies. Brown called SETI a valid science practiced by experts who expect to discover other intelligent beings in the universe.

Congressional advocates of SETI argued that it was not a futile hunt for little green men on Mars. They called attention to three of its strongest features: SETI's use of radio telescopes to monitor faint radio emissions amidst cosmic interference; its innovative use of computer technology to sort out incoming signals; and its role in stimulating young people to seek careers in science.

In 1992 NASA's SETI Microwave Observing Project, as it was called, underwent a critical name change. It was now the High Resolution Microwave Survey (HRMS), and NASA moved it from the Life Science Division to the Solar System Exploration Division. There it became part of (TOPS), the Towards Other Planetary Systems program. The goal of the TOPS program was the detection of other planetary systems.

The reasons for these changes are not entirely clear. Steven J. Dick, the best-known chronicler of the twentieth-century search for extraterrestrial intelligence, writes that the Senate Appropriations Subcomittee changed SETI to HRMS. A remark made by Jill Tarter, a principal NASA project scientist, seems to support Dick's interpretation of events. After NASA announced the name change, Tarter said that HRMS "is now the politically correct name for this exploratory endeavor."[5]

Some newspaper articles, however, claim that friends of NASA in Congress were responsible for the name change. These supporters hoped that HRMS would divert critics who equated SETI with the hunt for little green Martians. This account is reinforced by an article published in *Science*, "SETI Faces Uncertainty on Earth and in the Skies." Its author, Richard A. Kerr, saw the moves by NASA as a way to lower SETI's profile and avoid "future congressional potshots." Kerr believed that NASA's concealment of SETI behind the screen of name and division changes was a stopgap measure likely to fail.

Whatever the ultimate reasons for the change, SETI officially became HRMS. Overt references to the controversial search for extraterrestrial intelligence faded into the background in the new acronym. Instead of searching for signs of extraterrestrial intelligence, NASA was surveying the skies for high resolution microwaves.

NASA's High Resolution Microwave Survey began its search for intelligent life on Columbus Day, 1992, at 3 P.M. That date commemorated the 500th anniversary of the Italian navigator's discovery of the New World. The *New York Times* drew parallels between Columbus and astronomers who searched for new worlds using radio telescopes. NASA intended first to deploy antennae in Puerto Rico and California to detect these new worlds. Radio telescopes located in West Virginia, Australia, Argentina, Russia, and India were scheduled to join the NASA-sponsored hunt for intelligent life.

At the opening ceremony, John Billingham of NASA said: "We sail into the future, just as Columbus did on this day 500 years ago."[6] The comparison was not quite appropriate for the occasion. Columbus set sail in August of 1492, and by October 12 he had reached the Bahamas. On October 12, 1992, NASA's voyage of discovery had just begun. It had yet to contact new worlds in the sky.

Astronomers activated radio telescopes at the Arecibo site in Puerto Rico and in the Mojave Desert. The computer system designed to process multiple incoming transmissions was capable of searching millions of channels simultaneously. The aim was to detect any pattern that might suggest an intelligent message coming from the vicinity of a star.

As NASA personnel began their work, one investigator spoke enthusiastically about his hopes. Dr. Peter R. Backus, who was conducting tests on the signal processing system, noticed some peculiarities in several signals. Backus told a *New York Times* reporter that within a few weeks "we just might do it for real. Who knows?"[7]

NASA prepared itself for early contact with intelligent aliens. Urged on by John Billingham, leading international astronomical societies drafted a Declaration of Principles Concerning Activities Following the Detection of Extraterrestrial Intelligence. The declaration called for a careful verification of a likely message and then its prompt dissemination through scientific channels and the popular media.

Not all scientists believed that an extraterrestrial message would soon arrive on Earth. Harvard University's premier evolutionary biologist, Ernst Mayr, wrote a letter to *Science* in which he noted that physical scientists and engineers dominated HRMS. The critical dimensions of the problem they explored, he continued, were not physical. They were biological and sociological.

Mayr said that 50 billion species have lived on the Earth, and only one of them generated civilized life. Of the twenty some civilizations appearing in history, only one developed electronic technology. Mayr called the NASA search "highly dubious," an extravagant expenditure of money in times of appalling federal debt.

Congress apparently agreed with Mayr and other critics of the search for alien intelligence. In October 1993, one year after HRMS first set sail, Congress refused to renew funding for the project. At this point, NASA had already spent over $50 million to develop the sophisticated equipment needed for a continuous high resolution survey. Despite this expenditure, HRMS was not an expensive undertaking. During its lifetime, HRMS was expected to consume less than one-tenth of one percent of NASA's $14.6 billion budget.

In the Senate, Richard H. Bryan of Colorado was angry. He accused NASA bureaucrats of deliberately changing SETI to HRMS in order to hide the true aims of the program. Bryan called HRMS a waste of taxpayer dollars and persuaded all but twenty-three senators to join with him in eliminating funds for the program. Congress terminated HRMS and appropriated $1 million to shut it down. A newspaper columnist wrote that it was as if the Great Navigator had barely sailed beyond the Canary Islands when he received a message ordering him to return because Queen Isabella decided to keep her jewels. The tired clichés of the Columbus story persisted in the midst of SETI's worst public disaster.

There was more to come. In 1993 physicist Alan Cronmer published a book on the nature of science in which he criticized HRMS sharply. Cronmer called it the "space-age version of communicating with God"[8] and pointed to the religious fervor of the enterprise. At best, he said, SETI was fringe science in which researchers piled improbability upon improbability. He compared it to looking for the Loch Ness monster.

The end of HRMS was felt most sharply within the small SETI community. It was not a matter of major concern for scientists and scientific organizations at large. Scientists outside the SETI circle were generally indifferent to the end of NASA's attempt to search for signs of intelligent extraterrestrial life.

SETI Perseveres

The search for extraterrestrial intelligence continued after the termination of HRMS, but it no longer had official status as a NASA program. NASA lent prestige, authority, and credibility to an endeavor that operated on the outer limits of scientific respectability and Congressional funding. Eventually, SETI research was continued with private funds. Nevertheless, it was difficult to regain the position SETI and then HRMS held when they were under NASA sponsorship. The space agency was highly regarded by the public. NASA had sent men to the Moon, Viking landers to Mars, and spacecraft on voyages that would take them to the limits of the solar system and beyond. A private institution could not match that record, no matter how well it managed a renewed search for alien intelligence.

The burden of SETI research was assumed, in part, by the SETI Institute. Its members took the end of HRMS quietly. Astronomer Seth Shostak said, "No one's falling on their swords here."[9] Instead, the Institute immediately began a search for private funds to replace NASA money. Contributions soon arrived from wealthy SETI supporters. Contributors included science fiction writer Arthur C. Clarke, Paul Allen, who cofounded Microsoft, Gordon Moore of Intel, and David Packard and William Hewlett of Hewlett-Packard.

The SETI Institute took over a portion of the search begun by HRMS. The NASA project initially included an all-sky survey for incoming signals and a targeted search aimed at one thousand sun-like stars located within 200 light years of the Earth. The targeted search became Project Phoenix under the auspices of the SETI Institute. Project Phoenix was placed under the direction of Jill Tarter, who had worked on NASA's HRMS.

Projects at American universities, notably Ohio State, Harvard, and the University of California, Berkeley supplemented research done by the SETI Institute. Initially, American and Soviet scientists showed the greatest interest in collecting evidence of extraterrestrial life. Somewhat later researchers in Argentina, Australia, and Holland joined them. The rest of the world's scientists were noticeably absent from a venture that Carl Sagan often said held the greatest promise for the future of humanity. Of 7,000 professional astronomers working around the world, about 10 percent are concerned with SETI projects.

At the University of California's Berkeley campus, computer scientists finally found a way to engage a larger portion of the world's population in the search for alien intelligence. They created software for the famous SERENDIP SETI Project. This software uses idle time on home and office computers to download and analyze radio telescope data. By 2004 nearly 5 million participants from more than 200 countries agreed to process the flood of data collected by radio telescopes. The spectacular success of SETI@home demonstrated widespread popular interest in joining the search for extraterrestrial intelligence.

Berkeley scientists had created a huge, powerful virtual computer by harnessing the power of millions of personal computers. They gave it the job of searching radio noise for nonrandom patterns that indicated a signal from another civilization. In 2000, Frank Drake reported that the new search technology was 100 trillion times more powerful than the equipment he used at Green Bank in 1960. Despite this new capability, no radio astronomer has yet detected a legitimate extraterrestrial message.

Pond Scum

While computers around the world were processing incoming radio signals, and Project Phoenix was underway at the SETI Institute, NASA reentered the study of the nature and origins of terrestrial and extraterrestrial life. The new emphasis was on life in the universe, not on alien intelligent life as such.

In 1998, five years after the abrupt cancellation of HRMS by Congress, NASA cautiously and judiciously revived the study of extraterrestrial life. Now it was one of the subjects studied by the newly created Astrobiology Institute, located at NASA's Ames Research Center. The new Institute was placed under the directorship of Baruch Blumberg, winner of the 1976 Nobel Prize in physiology and medicine. Blumberg directed a "virtual" Institute because its teams of researchers, scattered around the world in universities and laboratories, communicated with one another electronically. On occasion they assembled for more traditional scientific meetings.

The international group of geologists, chemists, oceanographers, planetary scientists, molecular biologists, virologists, zoologists, and paleontologists who gathered at astrobiology meetings endorsed the goals of the new science: to study the origins of life on Earth, determine how it might have arisen elsewhere in the universe, and establish ways to locate and recognize it beyond the Earth.

For the most part, astrobiologists avoided the topic of alien intelligence that had undercut public funding to NASA in the recent past. At an early meeting of astrobiologists, Frank Drake said that intelligent creatures, not "pond scum," were the primary interest of those assembled. However, working astrobiologists initially appeared more interested in the "pond scum" side of their research.

Members of the Institute studied microorganisms that live at extreme temperatures (hot and cold) on Earth, examined meteorites for traces of extraterrestrial microbial life, and looked forward to searching soil samples from Mars or the icy ocean of Europa (a moon of Jupiter) for signs of life. Finally, they are interested in the possible detection of biological activity in the atmospheres of one of the many recently discovered extrasolar planets. The radio telescope search for extraterrestrial intelligence does not dominate these ventures as it did the old SETI projects.

NASA did not revive HRMS, but it began to alter its anti-SETI stance, a stance that dated to the 1993 termination of Congressional funding. Frank Drake said that for nearly a decade after Congress eliminated funds for NASA's search for alien intelligence, SETI was a four-letter word at NASA. It was not used in speeches and documents coming from the space agency. The opposition to SETI within NASA began to soften in the early twenty-first century when, as the *New York Times* reported, SETI research began to gain respect bit by bit.

The change of attitude at NASA was a result of a renewed scientific interest in the origins of life on Earth, questions about life on Mars, and the possibility of life on newly discovered extrasolar planets. NASA's change of attitude was also influenced by research conducted by the SETI Institute. The Institute continued to hunt for radio signals of extraterrestrial origins and study problems in the new science of astrobiology. Using private funding, and grants from NASA and the National Science Foundation, scientists at the Institute continued their investigations outside the confines of the space agency.

In the summer of 2003, NASA formally named SETI Institute scientists to one of twelve new teams in its Astrobiology Institute. Team leader Christopher

Chyba, a SETI Institute scientist and student of Carl Sagan, received a five-year NASA grant to study planetary biology, evolution, and the nature of intelligence. The funding was small, about one million dollars a year, but the symbolic value was large. SETI was back at NASA after an embarrassing defeat in Congress and a decade of exile.

CHAPTER TEN

Mirror Worlds

> The future of science won't be like the comforting picture painted in *Star Trek*: a universe populated by many humanoid races, with an advanced but essentially static science and technology. Instead, I think we will be on our own, but rapidly developing in biological and electronic complexity.
>
> —Steven Hawking, *The Universe in a Nutshell*, 2001

Universal Science?

As early as the sixth century B.C., Xenophanes criticized the Greeks for modeling their gods and goddesses after human beings. He satirically declared that if cows and horses had hands and could draw, they would model the bodies of their gods after themselves. And in the middle of the eighteenth century, the British philosopher David Hume observed there was a universal tendency among humans to conceive all beings like themselves, "and to transfer to every object, those qualities, with which they are familiarly acquainted."[1]

These sources recall the long history of anthropomorphic thought and its continuing influence in modern times. Despite the efforts of SETI scientists to avoid the pitfalls of anthropomorphism, they duplicate terrestrial life and civilization on distant planets, creating a succession of alien worlds that mirror their own.

SETI investigators tend to transfer terrestrial life and culture to the rest of the universe because they operate beyond the limits of their knowledge and competence when they discuss the universality of science and mathematics, biological and cultural evolution, the idea of progress, the nature of technology, and the meaning of civilization. Astronomers and physicists first meet these complex areas of knowledge when they venture into history, philosophy, and the biological and social sciences. Not surprisingly, they use concepts drawn from the physical sciences to determine the nature of alien cultures.

Searchers for extraterrestrial intelligence suppose that alien mathematics and science are essentially like ours. When physicist Edward Purcell wrote about communication with extraterrestrials in the 1960s, he asked rhetorically: "What can we talk about with our remote friends?" His immediate answer was: "We have a lot in common. We have mathematics in common, and physics, and astronomy. . . . We have chemistry in common, inorganic chemistry, that is."[2]

Purcell not only assumed that the physical sciences are practiced throughout the universe but that alien science is bound to harmonize with terrestrial science. These premises, crucial to the belief that we can communicate with advanced extraterrestrial civilizations, are riddled with philosophical difficulties.

In a speech on the nature of science delivered in 1989, Nobel laureate physicist Sheldon Glashow noted that the recently discovered rings of Neptune were evident to American, Russian, Japanese, and Ugandan astronomers alike. The existence of the rings did not depend upon the gender of the observers nor upon their ethnic, national, or cultural backgrounds.

The universal nature of science practiced on Earth led Glashow to extend human knowledge of the physical sciences to the rest of the universe. He maintained that intelligent aliens would eventually develop "the same logical system as we have to explain the structure of protons and the nature of supernovae."[3]

Glashow's attempt to establish a cosmic physical science is not well founded. American, Russian, Japanese, and Ugandan scientists are *Homo sapiens* trained within the confines and traditions of modern science. On the other hand, virtually all scientific commentators on the subject agree that intelligent aliens are not like humans. They are not replicas of *Homo sapiens* who happen to live on an extrasolar planet.

Glashow makes no distinction between science practiced by different human groups and science practiced by intelligent creatures living on other worlds. However, human and alien science differ because there are enormous discrepancies in the biological constitution, intellect, and sociocultural lives of the two sets of practitioners.

Glashow's fellow Nobel Prize winner Steven Weinberg proposed translation as a way to bridge the gap between terrestrial and extraterrestrial science. In

1996, Weinberg argued that if we translate the scientific works of intelligent aliens into our terms, we will learn "that we and they have discovered the same laws."[4] The difficulty here is translation, an act studied by modern philosophers. If we meet aliens, how can we determine if they have a language and practice science? To simplify matters, suppose we overcome these initial problems. We will then transform alien science into something we recognize as our kind of science. The result of this transformation process does not produce universal science. It produces a form of knowledge cast in the image of terrestrial science.

Barry Allen, who criticized Weinberg's views on extraterrestrial science, commented: "Weinberg knows no more about how aliens think than you or I do."[5] Weinberg agreed with Allen's comment but added that he never intended to depict the true nature of alien science. He merely presented an "illustrative prediction." Weinberg's illustrative prediction is based on the way physicists of different national origins on Earth accept the validity of the same set of physical laws. Thus, Weinberg resorts to the same analogy of a multicultural human science that Glashow offered earlier.

Shortly after NASA launched its ambitious new search for intelligent life in 1992, an editor of *Scientific American* asked Frank Drake how it was possible to communicate with advanced life in the universe. Intelligent aliens, Drake said, developed systems of mathematics, physics, and astronomy similar to those found on Earth. He believed that general relativity, quantum-field theory, and superstrings were already part of alien physics. An innate curiosity about nature and the need to better their lives, Drake continued, compel extraterrestrials to explain physical phenomena as we do.

When philosopher Nicholas Rescher was asked to comment on Drake's notion of alien science, he dismissed it as infinitely parochial. It was like saying that extraterrestrials share our legal or political system. Rescher was well qualified to examine Drake's claims. He had recently studied the anthropomorphic character of human science and how it related to alien science.

Rescher struck at the heart of the popular conception of alien science when he challenged the widely held view that there is only one natural world and a single science to explain it. He called this the one world, one science argument.

The physical universe is singular, Rescher agreed, but its interpreters are many and diverse. What we know about physical reality stems from our special biological and cognitive make-up and our unique cultural and social heritage and experiences. We have no reason to suppose that extraterrestrials share our peculiar biological attributes, social outlook, or cultural traditions. Human science, therefore, is incommensurable with extraterrestrial science. If extraterrestrials cultivate science, it will be their kind of science, not our kind. Alien science is a wholly different form of knowledge. It is not human science raised to a higher degree.

Rescher offered a compelling illustration of how human biology and our situation on Earth shaped our science. Astronomy as practiced by humans has been molded by the fact that we live on the surface of the Earth (not underwater), that we have eyes, and that the development of agriculture is linked to the seasonal positions of celestial objects.

Intelligent alien creatures living in an oceanic abyss might develop sophisticated hydrodynamics but fail to study the motion of heavenly bodies, investigate electromagnetic radiation, or build radio telescopes. Even if extraterrestrials are surface dwellers, their biological endowment will determine what they are able to sense, their ecological niche, what aspects of nature they exploit to satisfy their needs, their cultural heritage, which questions about nature they find interesting to ask.

Rescher acknowledges the existence of intelligent extraterrestrials who possess the ability to develop science and technology. He does not dispute the scientists' repeated claims (1) that there is a single scientifically knowable physical reality and (2) that aliens are not simply other humans inhabiting a different planet. After adopting these claims, he demolishes the idea of a universal science that serves as a common language in the universe.

Rescher maintains that wherever science exists in the universe, it will be localized. It will be the science of the creatures who have fashioned it. They will act according to their special physical constitution, environment, history, and needs. Hence, science diverges in the universe. It does not converge on the theories, concepts, and topics that happen to interest terrestrial researchers at this point in the history of the human intellect.

Rescher accepts the real world of the scientist and believes that science yields unique knowledge about the inherent structure of reality. Nevertheless, he refuses to equate human science with the science created by beings who are biologically distinct and who inhabit radically different physical, social, and cultural milieus.

Searchers for extraterrestrial intelligence overlook the fact that modern science is a mere four or five centuries old. It was not available throughout the more than 5-million-year history of hominids. Our early ancestors survived, multiplied, and spread over the Earth without the help of science. Modern science is a notable human achievement, but it is not an absolute necessity for the survival of our species. Since science has not powered the long history of humanity, why should we assume it is a form of knowledge found everywhere in the universe?

The Evolutionists on SETI

Three well-known evolutionary biologists—Theodosius Dobzhansky, George Gaylord Simpson, and Ernst Mayr—mounted strong attacks on notions of the origin and development of intelligent extraterrestrial life held by SETI investigators. Their criticism focused on four issues: first, the deterministic thinking of scientists who portray evolution as a fixed process with preprogrammed goals; second, the contingent nature of organic evolution—mutations and unpredictable ecological changes make the evolutionary process dependent upon a chain of random circumstances; third, the role of intelligence in the adaptation of organisms (scientists in the SETI circle take the emergence of high intelligence for granted; most evolutionists see high intelligence as a rare event in the history of life); fourth, the anthropomorphism that typifies thinking in physics and astronomy about alien life—despite protestations to the contrary, the physical scientist's view of advanced life retains key human characteristics.

In 1964 George Gaylord Simpson published an essay in *Science* entitled "The Nonprevalance of Humanoids." He was inspired to write about humanoids (human-like creatures) because of the various research programs on alien life sponsored by NASA, and encouraged by the National Academy of Sciences, in the 1960s. The study of exobiology, Simpson argues, might have official sanction, but it is a science without any evidence to support it. Exobiologists may think of themselves as biologists, but they tend to know more about physics, chemistry, and biochemistry than they do about evolutionary biology.

Simpson doubts that humans would recognize life forms not based on the carbon chemistry that fostered terrestrial life. Organisms with some other chemical and structural basis would not fit classificatory systems devised by Earth-biased observers. Although he raises these and other objections, Simpson thinks it reasonable to suppose that life defined by terrestrial criteria may exist beyond Earth. However, Simpson reminds his readers that this is pure speculation on his part. It is not a fact.

Simpson criticizes scientists who envision an evolutionary path that culminates in intelligent creatures similar to humans. Evolutionary history, he counters, is opportunistic and unpredictable. It does not move deterministically toward preestablished goals. Instead, evolution makes do with what happens to be available at a particular time and under a given set of circumstances.

Humans beings are no exception to this rule. *Homo sapiens* are the result of a 3-billion-year-old causal chain of events. That chain cannot be repeated on some other planet unless the planet has a history identical in every detail, including every moment of time, to the history of the Earth.

Simpson writes that it is extremely unlikely that anything remotely like humans inhabits the universe. If such creatures do exist, it is impossible for humans to communicate with them. The fundamental differences between terrestrial and extraterrestrial organisms prevent the exchange of information between them.

Extremely unlikely does not mean impossible, and Simpson admits that others have the right to dream that they are not alone in the universe. Dreams of alien intelligence, however, remain dreams. They may inspire science fiction or poetic reflection but not scientific research.

Simpson understands that his rational arguments will not persuade those who search for signs of intelligent extraterrestrial life. Their emotional commitment and self-interest, he says, hinder his chance of success. Nevertheless, if astronomers persist in searching for extraterrestrial intelligence, they should know that their hunt is a gamble with the worst odds in history. That is why the search for alien intelligence resembles a wild spree more than a sober scientific program.

Dobzhansky's appraisal of the problem of extraterrestrial life appeared a decade after the appearance of Simpson's essay. Dobzhansky begins his article by clarifying the distinction between the origins of life and its subsequent evolution. He notes that most biologists avoid commenting on extraterrestrial evolution. By contrast, cosmologists and exobiologists assume that the development of extraterrestrial life recapitulates the appearance of intelligent life on Earth. Hence, they conclude that creatures similar to humans have established flourishing technological civilizations throughout the universe.

Simpson and Dobzhansky presented their ideas of alien life during the early decades of the space age. Ernst Mayr, writing in several scientific periodicals in the 1990s, confronted NASA's SETI program and the conception of intelligence adopted by its researchers. He based his criticisms on a lifetime study of the science, history, and philosophy of organic evolution.

Mayr argues that the $100 million allotted to NASA for its decade-long SETI project is a waste of federal funds. Astronomers, physicists, and engineers, ignorant of the crucial biological and social components of their venture, advise the space agency on SETI projects. Therefore, NASA's search for messages from advanced civilizations is a flawed if not futile effort.

Mayr accepts the *probability* that life originated independently on extrasolar planets resembling the Earth. He says that it is *improbable* that extrasolar planets nurture intelligent life and that it is *highly improbable* that alien life has evolved advanced intelligence. Given his reservations, Mayr all but ruled out the possibility of extraterrestrial civilizations contacting Earth via radio signals. He considered the possibility of extraterrestrial organisms receiving human-generated radio signals directly through special sensory organs but rejected it.

Mayr dismisses the argument that intelligence ensures the successful adaptation of an organism to its surroundings. Nor does he believe that human-level intelligence is a premium property for any creature. Of the billions of species that have inhabited the Earth, only one developed civilized life, and only one civilization mastered electronic communication. Perhaps civilizations are rare because high levels of intelligence do not benefit organisms. Many so-called higher creatures have levels of intelligence lower than humans. These include apes, monkeys, dogs, cats, whales, dolphins, and birds. None of the above developed civilized life or established electronic communication. Nevertheless, they have succeeded in surviving and reproducing themselves.

Evolutionary biologists claim that each species confronts peculiar environmental conditions and that there is no single property, including intelligence, that insures a species' survival. Millions of species have used other strategies to adapt, survive, and reproduce. According to Mayr, physical scientists are driven by a single-minded determinism. They erroneously believe that intelligence was a necessity early in the history of life and that its adaptive value increased thereafter. He has a ready explanation for this kind of thinking. Human beings are dependent on, and proud of, their superior intelligence. Consequently, they assume that other creatures cannot get along without it. An anthropomorphic impulse drives their discussions of evolution and intelligence.

Mayr notes that human intelligence comes at a steep biological price. It requires a large brain and complex central nervous system plus the metabolism to maintain them. It also demands a long infancy with extended parental care. That is why large-brained *Homo sapiens* appeared less than 300,000 years ago even though the hominid line branched away from the apes five to seven million years earlier.

The billions of species that have lived on Earth without intelligence, or with a low level of intelligence, were not at a disadvantage. They evolved other adaptations to cope with their ongoing struggle for existence. These alternative adaptations evolved more readily and more widely than the high intelligence we admire in humans and confer upon extraterrestrial organisms.

Mayr concludes that intelligence is a fluke of history. It is not an inevitable or necessary consequence of the development of life. Intelligence is one of many ways organisms deal with their environment. It is not a special property driving evolution along a progressive path. Or, as evolutionary psychologist Steven Pinker said, "Evolution is about ends, not means; becoming smart is just one option."[6]

• • •

Not every evolutionary biologist is as critical of the search for extraterrestrial life as Dobzhansky, Simpson, and Mayr. Some evolutionists have offered SETI their qualified support. Four distinguished evolutionists signed Carl Sagan's 1982 petition endorsing increased funding for SETI research. One of the signers, David Raup, reacted to criticism of SETI by proposing an organismic source for extraterrestrial signals.

Raup reviewed the arguments made by SETI enthusiasts, including claims that intelligent extraterrestrial creatures practice advanced science and technology and that they build and deploy electronic communications instruments. Raup agrees that evolutionary biologists have good reason to question such claims.

SETI investigators assume that intelligent extraterrestrial organisms build radio transmitters. Raup asks if creatures with nonconscious intelligence might transmit radio waves in some other fashion. We know that certain terrestrial organisms can detect magnetic fields and generate strong electrical currents. Electric eels and fishes, for instance, generate electrical fields that they use for seeking food and communicating with other members of their species. They generate electricity biologically, not technologically with dynamos. Likewise, some alien creatures might generate electromagnetic waves biologically, not technologically with radio transmitters.

Raup states that as late as 1991, biologists have found no living thing that transmits electromagnetic waves. Researchers in the future, however, might discover such a creature on Earth. A terrestrial organism able to generate radio signals could serve as a model for extraterrestrial organisms who have neither the conscious intelligence nor the manipulative ability to construct electronic devices.

Raup supports NASA's SETI projects because he believes if they succeed, humans will gain enormous benefits. He has difficulty, however, explaining how radio signals of biological origin will benefit us because the transmitting organisms are not necessarily intelligent. They have simply evolved the ability to send radio signals. And, how can radio astronomers located on Earth know where to search for electronic signals generated organically? SETI investigators claim that intelligent alien communicators deliberately choose radio frequencies based on their knowledge of the physical sciences. The same does not hold true for low-intelligence, biological transmitters.

Raup asks that we search for signs of incoming radio signals from organisms that have a minimum level of intelligence. Two-way communication is unlikely to take place under these circumstances. The alien signalers may accidentally reveal their existence, but they are not able to send coded messages or extensive information to Earth.

Another evolutionary biologist, Stephen Jay Gould, rose to defend SETI in 1982. Gould, a well-known popularizer of biology and evolution, approached the issue of intelligent extraterrestrial life cautiously. He ultimately endorsed SETI research because it was relatively cheap, promised great changes in human thought if successful, and did not contradict the theory of organic evolution. In supporting his third reason, he closely analyzed the response of evolutionary biologists to the idea of intelligent life in the universe.

Gould drew a distinction between the *specific* and *general* claims made by SETI practitioners. The specific claim, which most evolutionary biologists including Gould discount, calls for the near-exact repeatability of long sequences of evolutionary events. In this case, it means that evolution operating on a distant planet will produce creatures resembling humans. Gould argued that if the evolution of life on Earth were started anew, it would not necessarily end with the appearance of *Homo sapiens*. Using an analogy taken from magnetic tape recording, he said that if the tape of life were run through once again, the results would not be the same.

Most evolutionists reject the specific interpretation of organic evolution because they believe that evolution is a complex process filled with historical accidents along the way. Gould listed two major objections to the specific argument. The first is the mass extinction of organisms in the past. An asteroid happened to strike the Earth 65 million years ago. Dinosaurs, who lived on Earth for 140 million years, became extinct and opened the way for the evolutionary development of mammals. This cataclysmic event, which eliminated a dominant form of life, underscored the random nature of evolution.

Gould's second argument emphasized the contingent nature of the evolutionary process. The evolutionary chain of any species extends into a past filled with chance interactions between species and species, and environment and species. Evolutionary paths shift in one direction and then another again and again. According to Gould, any species is the result of a series of unique happenings. Its history is not repeatable on Earth nor on another habitable planet.

Gould could not defend the specific claims made by SETI scientists. However, he accepted the looser claim that intelligence in some unspecified or unimaginable form might exist elsewhere in the universe.

Gould thought it was possible for exotic alien life forms to converge on intelligence. On the Earth, convergence was evident in the separate evolution of flight in insects, birds, and bats. If convergence resulted in flight appearing in species belonging to different lineages, perhaps convergence might lead to the emergence of intelligence in extraterrestrial life forms. Gould was satisfied with this argument for extraterrestrial intelligence and believed it acceptable to other evolutionists.

Despite his use of convergence to bolster the existence of alien intelligence, Gould had low expectations for SETI's chances of success. He said that the probability of alien contact is much lower than that calculated by optimistic physical scientists. Nevertheless, the whole venture is worth a try. The curiosity that drives humans may also drive intelligent beings inhabiting other parts of the universe.

Gould's endorsement of the general claim for extraterrestrial intelligence rests upon his belief that these creatures are not similar to humans. Intelligence can appear in alien life forms with different anatomical structures. They might be blobs, films, spheres, masses of pulsating energy, or even more diffuse and unimagined shapes.

Gould and other evolutionists might settle for the general claim that intelligent life in some unspecified form may inhabit the universe. Living blobs, spheres, and films, however, are not suitable candidates for constructing and operating radio transmitters. Technology, as we know it, probably resulted from a combination of a big brain capable of comprehending the physical world and manual dexterity enabling an organism to manipulate it. In short, advanced terrestrial technology is a unique product of humans.

A small number of biologists believe that there is a limit to the number of evolutionary possibilities. One of this group wrote in 1964: if we succeed in communicating with extraterrestrials "they won't be spheres, pyramids, cubes, or pancakes," "they will look an awful lot like us."[7]

The British paleobiologist Simon Conway Morris revived this argument in a book he published in 2003. Specifically, Conway Morris disputed Gould's claim that evolution was the result of a series of random events. If the tape of life were rerun, said Gould, it would yield a different set of life forms that did not include humans. In criticizing Gould's contingent evolution, Conway Morris presented his views on the constraints limiting the direction of evolution, the nature of intelligence, and the existence of extraterrestrial beings.

At the heart of Conway Morris's argument is convergence and the repeated emergence of complex biological systems. All evolutionary biologists acknowledge that convergence plays a role in the evolution of life. They agree that different species, living under similar environmental conditions, can independently evolve similar characteristics. Eyes, for example, have evolved in unrelated species a number of times.

Conway Morris uses examples of convergence drawn from a wide variety of sources to expand the role of convergence in shaping evolution. In his world, evolution is confined to a limited number of paths because species tend to converge on the same solutions to produce similar body plans and biological mechanisms. Therefore, when Conway Morris reruns Gould's tape of life, he expects it to produce creatures much like us. These creatures might show some slight differences from humans, but nothing of importance.

Likewise, evolution operating on another planet will produce extraterrestrial beings that resemble humans. And, since Conway Morris is willing to extend his thesis beyond biology, extraterrestrial cultures will converge upon agriculture and tool use as practiced on Earth.

At first glance, Conway Morris appears to support the views of life and culture held by SETI researchers for years. However, he makes it clear that although extraterrestrial life and culture would reflect its terrestrial counterparts, life does not exist beyond the Earth. Hence, he argues, "Life may be a universal principle, but we can still be alone."[8] Conway Morris is not impressed by attempts to create life in the laboratory. He is convinced that the initial appearance of life was due to a set of extraordinary circumstances not easily repeated on Earth or elsewhere in the universe.

Conway Morris's conclusions on these matters are probably influenced by his personal religious beliefs. Remarks supporting a religious outlook and critical of materialistic interpretations of life run throughout his book. The last part of his work is entitled "Towards a theology of evolution." There he announces his belief that Darwinism and religion are compatible. Given Conway Morris's search for a common ground between science and religion, he believes that the teleological approach has a rightful place in the search for scientific truths.

According to Conway Morris, the many convergences guide evolution along a progressive path that leads to intelligent, human-like creatures. Living things are not products of a helter-skelter process. Instead, long-term trends constrain evolution to a goal of complex human-like creatures. If humans recklessly destroy themselves, there are other intelligent species waiting in the wings to follow the converging paths that end in high intelligence, culture, and tool use.

All of the above takes place on Earth, not in the heavens. Conway Morris's final message is summarized in the subtitle of his book, *Inevitable Humans in a Lonely Universe*. The universe, as he sees it, is not without purpose or plan. Nor does it lack a creator who is the lord of all creation.

Progress

Unlike many evolutionists, Conway Morris finds evidence of progress in the evolution of life. The emergence of complex animals with larger brains with the ability to communicate with one another, live in advanced social systems, and create culture is proof enough that progress has occurred. His claim that life converged upon agriculture and tool use goes beyond biology into the realm of culture, where it raises questions that have long troubled historians, philosophers, and anthropologists. The idea of progress, the meaning of technology, and the nature of civilization are often mixed together in discussions of progressive technological civilizations. These complex subjects deserve special attention. They cannot be understood by lifting concepts and theories from the biological or physical sciences.

Scientists searching for extraterrestrial intelligence often cite goal-oriented progress as proof that human evolution is progressive. The goal of organic evolution becomes the production of intelligent creatures able to produce sophisticated technologies. It is possible, however, to choose other goals with different results. For example, we can define the aim of evolution as the domination of the terrestrial biomass, the total mass of all living things on Earth. By many measures—species longevity, total biomass, ability to cope with widespread catastrophes—bacteria easily win the contest.

Forests, often considered the largest component of the biomass, contribute far less to the total biomass than bacteria. One scientist estimates that bacteria living beneath the Earth's surface account for 2×10^{14} tons of the biomass. This figure exceeds the mass of all flora and fauna living on the face of the Earth.

The lineage of bacteria extends back more than 3.5 billion years while our earliest human ancestors first appeared 5 to 8 million years ago. And the germ theory of disease demonstrates that bacteria can cause humans to sicken and even die. This does not prove that bacteria rule the Earth or that they are superior to humans. It does raise questions about how to define progress and direction in the evolution of life on Earth or elsewhere in the universe.

The idea of progress, a creation of early modern Europe, has few roots in antiquity or the Middle Ages. Its origins are evident in its strong Eurocentric bias. Western civilization is the standard by which the progressive achievements of all other cultures are judged. The idea of progress reached its high point in the early twentieth century. Since then it has come under attack from critics who point to a variety of persistent problems that undermine a simple faith in human progress. How can we celebrate progress, the critics ask, in an age threatened by overpopulation, intractable diseases, environmental pollution, wars, terrorism, religious conflicts, and the widening gap between the rich and poor?

Throughout its history, progress has carried at least two meanings. It can mean forward movement toward a stated goal, or more broadly, the betterment of the human condition. In practice the two meanings are often merged, and the goal of progress becomes the advancement of humanity. Modern writers tend to stress human advancement in scientific or technological terms rather than in moral or cultural ones. Thus, technological innovations and scientific discoveries serve as convenient markers for the progress achieved by a particular civilization, nation, ethnic group, race, and so on.

Technology

Seekers of extraterrestrial intelligence have adopted some mistaken notions about the nature of technology. They assume that technology moves progressively toward goals predetermined by the universal laws of science. The pathway of technological development culminates in interstellar communication by space ships, probes, or radio waves.

The Project Cyclops report of 1971 argued that despite differences in intelligent life forms, their technologies converge. At some point in the history of extraterrestrial technology, the report announced, "microscopes, telescopes, communication systems, and power plants" must be similar to ours because they are based on the same physical principles. The report claims that technological systems are products of the laws of optics, thermodynamics, and theories of electromagnetism and atomic reaction and not the peculiar attributes of the creatures who happen to design them. Because of the universality of science, mathematics, and technology, communication with extraterrestrial beings is assured.

The Cyclops report does not consider the influence of cultural factors on the development of technology. Given our knowledge of the history and philosophy of technology, we know that our technology could have developed in many different directions. Science is a necessary, but not sufficient, condition for the production of technology. Granted, modern technology owes much to the growth of scientific knowledge. However, there are important technologies that are not dependent upon modern science. The making of a wide array of tools and weapons from stone and plant material persisted without the help of science for several million years. The controlled use of fire appears long before the rise of science.

Technology produced by the application of modern terrestrial science is constrained by the nature of that science. Modern technology is shaped by the ways humans have constructed their view of the physical universe. The physical universe limits that construction, but it does not absolutely define it. Just as there is no universal science, there is no universal technology.

Popular accounts of the history of technology claim that the stage of interstellar communication is reached through a well-defined sequence of technological events. They include the use of stone tools by our early human ancestors, origination of language, discovery of fire, emergence of ceramics and metallurgy, development of agriculture and sedentary living, invention of writing, cultivation of early mathematics and astronomy, rise of modern science founded on observation and experimentation, and creation of mechanically based industries. All of this reached its high point in the establishment of electronic technology, a technology dependent on the application of modern science to communication technology.

Historians have no proof that technology follows this or any other predictable sequence of stages. If a technological tradition begins with stone hand tools, it need not end in electronic communications. The history of technology is filled with technological paths never followed. Once a particular technology is developed, and social, cultural, and economic commitments made to it, then other technological possibilities are closed. The opening of a door to one technological solution closes off outlets to its alternatives.

So the history of technology does not follow a single path that leads from stone tools to radio telescopes, or any other technological process or artifact. A more fruitful analogy is a many-branched bush with some technological branches fully developed while other branches are left unexplored, or partially explored and abandoned. The historical record demonstrates that humans have lived in radically different technological settings. There is no single technological way of living as a human being. Over time, different societies, using different technologies, have survived and flourished. We tend to overstate the influence of technology on human survival. Even controlled fire arrived late on the scene, perhaps 250,000 years ago or earlier.

John Ball, who originated the zoo hypothesis to explain the absence of alien visitors to Earth, also offered an analysis of the overall evolution of a technological civilization. There were, he claimed, three possibilities. First, a civilization could be destroyed by technology externally or internally. Second, a technological civilization could stagnate, showing no signs of progressive development. Civilizations with a low level of technology, he continued, "would eventually be engulfed and destroyed, tamed, or perhaps assimilated."[9] Thus, we are left with the third possibility, a civilization that shows quasi-technological progress, where progress is defined as control over the environment. These progressive, advanced civilizations have mastered the universe and are the only ones of interest. Aliens who control the universe may act as the zookeepers of other intelligent creatures who have not reached the technological prowess of their overseers.

Kardashev's three tiers of alien civilization is yet another way of thinking about alien technology in terms of evolutionary progress. The Soviet astronomer accepts the idea that as a culture uses more energy, it progresses to a higher level of civilization. A Type III galactic super-civilization has access to far more energy than a terrestrial industrial civilization; hence, it is superior to it. This is not a new idea. In 1928 author Aldous Huxley poked fun at thinkers who claimed that because humans now use 110 times more coal than their ancestors, they are 110 times more civilized.

The viewpoint Huxley satirized in the 1920s was at least a century old. It first appeared in the early days of the Industrial Revolution when steam engines were equated with the progress of British civilization. A number of nineteenth-century writers believed that excess energy made available by steam engines advanced the level of civilized life in Great Britain.

The equation of energy and civilization was periodically revived thereafter. In the early twentieth century, it was used by physical scientists to promise a paradise on Earth based on free energy from reactions at the atomic level. This promise was renewed after World War II when scientists and laypersons alike imagined a utopian world filled with automobiles, airplanes, and ocean-going ships powered by nuclear reactors.

The weakness of the energy-civilization equation is evident when we ask how the surplus energy delivered by steam engines or nuclear power plants is used by society. The additional energy can serve socially constructive purposes or be wasted on the production of trivial goods or warfare. Were the Soviets and Americans more civilized than other nations when they stockpiled enormous quantities of nuclear energy in missiles aimed at each other's cities? Measured in the quantity of energy per capita, they controlled more energy than any other people on Earth. The coupling of high energy use with civilization illustrates the defects in the notion that advanced technology acts as a civilizing force throughout the universe.

The idea of a progressive technological civilization is one of the weakest links in the chain of arguments used by searchers for extraterrestrial intelligence. Civilization, like progress, is a latecomer to Western thought. And civilization, like progress, is a vague term burdened with value judgments.

Civilization

"Civilization" became a popular term in the eighteenth century when it defined a polished and refined state of society. Civilization was contrasted with barbarism or savagery, which possessed much lower levels of social organization, moral behavior, artistic sensibility, and knowledge. Many nineteenth-century anthropologists mistakenly believed that all human societies pass through a savage and barbaric stage before they reach the heights of civilized societies exemplified in Western Europe.

By the nineteenth century, science and technology became important parts of the definition of civilized life. The existence of science and industry in

Europe and America was proof of the superiority of Western civilization. Modern searchers for intelligent extraterrestrial life consciously chose advanced technological civilizations as their models for alien societies.

The average lifetime (L) of an extraterrestrial civilization is an important element in Drake's equation. If L is large, then the number of communicating extraterrestrial civilizations (N) is correspondingly large. Short-lived civilizations appear and quickly disappear before they have a chance to communicate with Earth. Long-lived civilizations have a better chance to develop technologies to the point where they can communicate by radio with humans. Estimates of L range from ten thousand to one hundred million years with one to ten million years the figures most often cited. Given that range, humans are more likely to encounter alien civilizations that originated several million years ago.

When Carl Sagan spoke to the television audience of *Cosmos*, he extended the longevity of extraterrestrial civilizations. If a civilization manages to control high technology, and if its members avoid self-destruction, then the long-term future of the civilization is assured. Sagan claimed that the time scales of such civilizations approach periods of geological change or stellar evolution. Measured by these standards, alien civilizations might endure for hundreds of millions or billions of years. In that case, our Galaxy is filled with millions of advanced civilizations.

Sagan and his fellow searchers for advanced technical societies among the stars seem to misunderstand the nature of the terrestrial civilizations they have chosen to place on distant planets. They wrongly assume that civilized societies are structurally stable social systems capable of enduring for vast periods of time. Civilization, according to this viewpoint, is the final product of progressive cultural evolution. If its members evade disease, devastating wars, resource waste, and overpopulation, then there is virtually no end in sight.

This picture of civilization is simply wrong. Civilized societies on Earth are unstable social entities with relatively brief lifetimes. The words we use to describe how civilizations develop emphasize their short trajectory, not their enormous longevity. Civilizations initially rise, reach a high point, begin to decline, and eventually collapse.

The rise, decline, and fall of the Roman Empire is the best known example of the course of a civilization. In the eighteenth century, it inspired Edward Gibbon to write his classic history of Roman civilization. There are, however, other well-known civilized societies. Egyptian, Minoan, Mycenaean, Hittite, Mesopotamian, and Mayan civilizations, to name a few. Each of these reached great heights only to collapse and disappear. Civilizations on Earth do not have long life spans. At most they last for a few thousand years, not for hundreds of thousands or millions of years.

When civilizations fall, they leave behind material remains for poets to ponder and archaeologists to study. The ruins of past civilizations litter many of the continents on Earth. Biblical prophets and religious leaders called attention to the scattered remnants of vanished civilizations. Somewhat later, European poets of the Romantic era found melancholy pleasure in contemplating the ruins of once great civilizations. Both prophets and poets were inspired by what humans in the past had achieved and lost (Fig. 10.1).

Although he wrote enthusiastically about million-year-old alien civilizations, Carl Sagan knew that terrestrial civilizations do not last. His ideas of civilized life came from two very different sources. The first source was his study of world history and its failed civilizations. Sagan wrote specifically about the course of Mesopotamian, Greek, Roman, and Aztec civilization.

FIG. 10.1. A figure contemplating the ruins of a once great civilization. (C.-F. Volney, *The Ruins, or a Survey of the Revolutions of Empire*. London, 1822. Courtesy Special Collections, University of Delaware Library. Newark.)

The other source of Sagan's views on civilization was science fiction, notably the Martian novels of Edgar Rice Burroughs. From Burroughs he learned that Barsoom was a dying world in which barbarous green Barsoomians threatened the survivors of an ancient civilized race. The ancient Barsoomians had ruled the planet for over a million years. Proof of their glorious past was still visible in the deserted, decaying buildings spread across the desert planet.

With the ruins of antiquity and Barsoom in the back of his mind in 1966, Sagan asked: "Are the sands of Mars today drifting over the edifices and monuments of an ancient civilization?"[10] The answer, he said, must await the exploration of the planet by its first human visitors. Those visitors might uncover a civilization as much as 100 million years old.

Ten years after Sagan made these remarks, NASA's Viking program sent automated laboratories to analyze Martian soil. The instruments dispatched to the planet found neither living organisms nor organic molecules there. Despite these negative findings, Sagan was not ready to accept a lifeless Mars. Instead, he obtained a NASA grant to study Viking orbiter images of the Martian landscape in hopes of finding signs of an advanced civilization. He had his student assistants scrutinize the images for the Martian equivalents of the Great Wall of China, the Inca road system, or Roman viaducts. His assistants found no ruins of an ancient Martian civilization in the dust that covered the planet.

The leap of extrapolation from short-lived terrestrial civilizations to million-year-old extraterrestrial civilizations highlights the reasoning that lay beneath Drake's equation and the search for intelligent aliens. The impulse behind this leap is the necessity to produce a large N, a substantial number of advanced communicating civilizations in the Galaxy. In order to reach this goal, scientists assume that short-lived terrestrial-type civilizations attain very long life spans elsewhere in the cosmos. They believe that once current terrestrial difficulties are overcome, civilized life can sustain itself for very long periods of time.

Examples of long-term social systems exist on Earth, but they are not civilizations, and they are not featured in discussions of intelligent alien life. Tool-using hominids originated in Africa about 2.8 million years ago. Small bands of foraging hominids left Africa 1.7 million years ago for Europe and Asia. Gathering wild plant foods, scavenging meat, fishing, and hunting small animals, these foragers eventually spread over the Earth. Although they domesticated neither plants nor animals, foragers represent the oldest and most successful hominid adaptation.

The forager mode of life preceded agriculture, the use of metals and ceramics, and the founding of cities. Bands of foragers flourished long before the rise of civilization. Their technology featured tools manufactured from stone, wood, and other plant material. And, they eventually learned how to control the use of fire. The simplicity of forager material culture is deceptive. The foraging life includes a sophisticated knowledge of the location of food and water, animal behavior, growth cycles, variations in the seasons, landscape and landforms, and the specific use of plants and animals as food, medicine, and raw material for tool making.

Despite the knowledge they accumulated and depended upon for survival, early foraging bands did not practice modern science. Yet the social and economic life practiced by our earliest ancestors lasted many times longer than civilized societies that built great cities. The pressing questions are whether modern civilization will endure, given the poor performances of past civilizations, and why some scientists believe very old civilizations exist in outer space when terrestrial civilizations have short life spans.

There are a number of reasons why civilizations decay and fail. These include disease, climatic changes, internal strife, powerful intruders, and political and economic mismanagement. Experts argue over the cause of decline in specific instances, but there is one feature that all civilizations share and that may account for their fragility. Unlike the social organizations of foraging peoples, civilizations are noted for their complexity. Complexity here includes the size of civilized societies, diverse social roles assigned to members, the varied personalities within social groups, and the hierarchical status of social arrangements and governance. The economic and political structures erected by civilized societies are likewise very complicated.

Most modern humans live in complex societies and think that their way of life is normal and that it existed in the distant past. In fact, civilized societies are an anomaly. Anthropologist Robert Carneiro estimated that 99.8 percent of human history was dominated by small independent bands of foraging people. Complex societies, in the form of civilizations, have been with us for about six thousand years. Once they are established, complex societies have a tendency to expand and dominate the peoples of the Earth. That was true of Roman civilization and is true of modern Western civilization.

Despite the apparent success of civilizations, their complexity makes them vulnerable to collapse. At a critical point, they dissolve into smaller units and lose their power of expansion. Other forms of social organization, with different political, social, economic, and technological bases, eventually replace them.

According to modern social theorists, civilizations are complex systems, and they are therefore prone to collapse into simpler ones. Complex societies do not collapse into chaos; they fall back into a state of lower complexity (Fig. 10.2).

The simpler the social system, the longer it is likely to endure. Thus, the idea of million-year-old extraterrestrial civilizations runs counter to historical evidence of terrestrial civilizations and the behavior of complex systems. Searchers for extraterrestrial civilizations have not explained why the social systems they have chosen to place in interstellar space can escape the failure inherent in

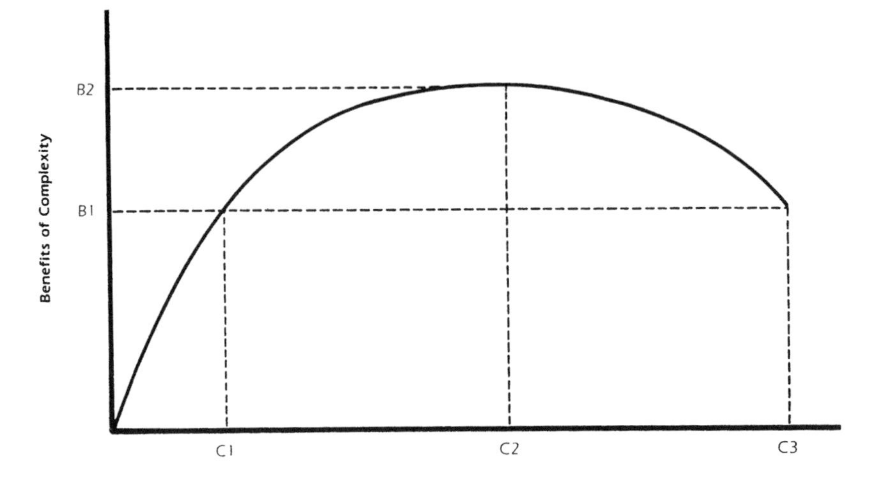

FIG. 10.2. Graph plotting the benefits of complexity to a civilization. A complex civilization reaches it zenith at C2 and then it begins to decline. (Joseph Tainter, *The Collapse of Complex Societies*. New York: Cambridge University Press, 1988. Reprinted with the permission of Cambridge University Press.)

complex terrestrial societies. They have little to say about the nature of extraterrestrial civilizations beyond the assertion that many of them have found solutions to problems troubling human societies and that they will persist almost indefinitely.

A terrestrial civilization may rise and decline over a short period, but what happens to its technology? Does it disappear with the civilization that originated it? Must each new civilization on Earth start from scratch and invent the technologies it needs to survive? The answer is "No" to each of these questions.

Although civilizations may come and go, their technologies persist. The story of fire technology, from its confinement in a circle of stones to central heating in modern buildings, is a continuous one. The culture that first gained control of fire is gone, and so are a number of cultures that used fire before our time. Nevertheless, fire technology persists. This is not an argument for technological progress but for the persistence of technology during a series of cultural changes.

Modern engineers are unable to replicate a small number of ancient techniques. They cannot build pyramids or cathedrals precisely as they were constructed in the past. However, these same engineers, using other means, can build pyramids and cathedrals if they choose to do so. There are no million-year-old civilizations on Earth, but there are million-year-old technologies here. The technology behind the shaping of stone tools began almost 3 million years ago and lasted into the twentieth century in New Guinea, Australia, and South

America. In Europe and America, modern anthropologists have learned how to make stone tools by replicating ancient stone-working techniques.

Does this mean that SETI supporters are correct when they claim that the lifetimes of some extraterrestrial civilizations extend over tens of millions of years? Their claims assume that the alien civilizations persist uninterrupted for very long periods, and that their technology develops progressively from level to level until it reaches stages far beyond those found on Earth. Without doubt, technology has continued on Earth despite its periodic association with declining or vanished civilizations. Nevertheless, there is no reason to suppose that all technology must develop along the paths terrestrial technology happened to follow after the decline of the civilizations that practiced them. Technology, far more than science, is limited and shaped by social and cultural factors. These factors arise from the unique history of the human race. That is why the history of terrestrial technology cannot serve as a template for the development of technology elsewhere in the universe.

Not only is terrestrial technology tied to human history; the very existence of technology in any form is questionable. A fair number of species on Earth practice technology as witnessed by spider webs, bird nests, rodent burrows, beaver dams, and chimpanzee tool use. There are other terrestrial species who appear to have little need for technology, or at least practice it at a very low level. On other worlds, with different sets of life forms, there may be no need to pursue technology to an advanced stage, or to practice it at all. Our tendency to populate extrasolar planets with Earthlike technology is more likely a sign of the failure of our imagination to conceive other possible ways of life than it is the likelihood that radio communication is a technological phase bound to emerge wherever there is intelligent life.

Granted, given an infinite universe, filled with infinite possibilities, and an infinite combination of those possibilities, radio technology similar to ours is bound to appear somewhere. However, if all possibilities have been realized, then intelligent extraterrestrials from other parts of the universe are living among us at the moment. To date, we have failed to detect these extraterrestrial visitors on Earth or in the heavens.

• • •

Many SETI scientists conclude that alien societies are little more than advanced copies of modern extraterrestrial civilization. They arrive at this conclusion because they are not aware of the nature of complex societies and because they project their immediate terrestrial experience into the universe. If extraterrestrial societies exist, they are not simply million-year-old versions of the industrial civilizations that currently flourish on Earth. Knowledge of the many ways life organizes itself socially on Earth is a first step to understanding that if creatures exist on other worlds, they may, or may not, organize themselves into complex societies. And if they do, their societies are not necessarily mirror images of the social organization of intelligent life on Earth nor of the technology it has developed over millions of years.

CHAPTER ELEVEN

Afterword

> We have found a strange foot-print on the shores of the unknown. We have devised profound theories, one after another, to account for its origin. At last, we have succeeded in reconstructing the creature that made the footprint. And Lo! it is our own.
>
> —A. S. Eddington, *Space, Time and Gravitation*, 1921

Two powerful strands run through the scientific search for extraterrestrial intelligence. The first strand is religion. There is religious sanction for populating the heavens with superior beings. The second strand is anthropomorphism. This is the tendency to describe the intellectual and social lives of those beings in human terms.

The religious strand is more obvious to outside observers than to most scientists investigating extraterrestrial intelligence. For centuries scientists have avoided offering religious solutions to scientific problems. Religion may have disappeared from formal scientific discourse, but the idea of superior celestial beings continues to influence scientific thinking about extraterrestrial intelligence. Robert Plank noted the psychological basis of the belief in advanced extraterrestrial life. He observed that humans have an emotional need to believe in the existence of superior celestial life. That need can be met by supernatural or natural beings.

Some scientists acknowledge the religious impulse that inspired their study of alien life. Carl Sagan was not among them. Although he was reluctant to accept religious interpretations of SETI research, Sagan used religious explanations to challenge the validity of UFOs. In an essay he contributed to the *Encyclopedia Americana*, Sagan argued that unidentified flying objects had less to do with scientific curiosity than with unfulfilled religious needs. For some people, he wrote, flying saucers "replace the gods that science has deposed."[1]

When novelist Cynthia Ozick interviewed Sagan for a popular magazine, she noted the religious overtones of Sagan's extraterrestrials. She said, "What you postulate is Angels. Faith, the same old faith." Sagan retorted, "Not faith. Calculation. Extrapolation."[2] Nevertheless, Sagan's eldest son, Dorion, supported Ozick's conclusions. Dorion, a science writer, scoffed at the search for extraterrestrial intelligence. He said it was nothing more than a replacement for religion in a secular age.

Meanwhile, Sagan's colleague Frank Drake readily admitted that a fundamentalist religious upbringing was the initial inspiration for him and others to join the search for advanced alien life. Granted, Drake repudiated his religious training in his youth. Nevertheless, in his mature years, he claimed that the extraterrestrials are virtually immortal and perhaps willing to pass on the secret of immortality to humans.

When depicting the life and culture of intelligent aliens, anthropomorphic thinking enters to fill in the missing details. Extraterrestrials, we learn, are remarkably like us. They study mathematics and science, practice technology, and grapple with issues raised by warfare, environmental pollution, diminishing natural resources, disease, overpopulation, and energy crises. This blatantly anthropomorphic portrayal of alien culture is accompanied by the disclaimer that aliens are biologically different from humans.

Anthropomorphic thinking, however, lies close to the surface of speculation about extraterrestrial intelligence. SETI pioneer Frank Drake was once asked what form an intelligent alien might assume. He answered:

> They won't be too much different from us. What I usually say, when people ask me that question, is that a large fraction will have such an anatomy that if you saw them from a distance of a hundred yards in the twilight you might think they were human.[3]

Drake continued his description by noting that it was advantageous to walk upright on two legs with your head on top, eyes near the brain, and mouth near the eyes. He clearly believed that human anatomy serves as a pattern for intelligent aliens. When Drake was asked if aliens were real, he candidly responded: "You talk about something enough times, you begin to believe it. And we sure talk about this a lot."[4]

Since the seventeenth century, philosophers and scientists have understood that anthropomorphic thinking exists in science. That is why scientists differentiate fact from value, resist granting humans a unique status in the universe, and avoid searching for purpose in the cosmos. Scientists claim they can produce objective knowledge even though their work rests upon human perception and understanding of the physical world.

Sagan had his own solution to the problems introduced by anthropomorphism. He asked scientists to reject the chauvinisms that marred their thinking. A chauvinism is simpler to overcome than entrenched anthropomorphic thought because the former is easier to recognize and vulnerable to rational argument.

The main chauvinism that threatens the study of extraterrestrial organisms is the widespread belief that life must have the same physical basis everywhere in the universe. Sagan identified oxygen, carbon, ultraviolet light, and temperature chauvinisms. Each of these rests upon the false assumption that terrestrial and alien life have identical chemical and physical requirements.

Sagan thought that chauvinisms are temporary hindrances to clear thinking that scientists can remove by critical analysis. In *The Cosmic Connection* (1973), he listed the prime chauvinisms about alien life, analyzed their flaws, and then moved on to discuss interstellar organisms. These highly intelligent creatures, he added, inhabit interstellar space and use novel technologies to convert the matter and energy of stars and galaxies to meet their needs. In effect, Sagan substituted one set of chauvinisms for another one.

Sagan criticized anthropomorphic thinking because it limited our understanding of extraterrestrial life. Despite these criticisms, Sagan and his SETI colleagues assumed that aliens build electronic equipment to communicate with others in the universe. Sagan's electronic chauvinism is far more crucial than the chauvinisms he exposed. The transmission and reception of radio signals over great distances is at the heart of SETI research projects.

Anthropomorphic thinking is more deeply embedded in science than most scientists realize. Philosophers claim that basic concepts of physics, such as force, attraction, and resistance, originated in human sensory experience. In the biological sciences, the nature and social arrangements of organisms are often viewed in the light of the species biologists know best—*Homo sapiens*.

Science is a human activity practiced within a given social and cultural context. Therefore, anthropomorphism is an irreducible part of scientific thought.

Scientists cannot escape the biases introduced by anthropomorphism. However, they have struggled to avoid them and obtain reliable knowledge about the workings of the physical world. Although anthropomorphism is not fundamental to science itself, and scientists are aware of its dangers, it persists as a part of science.

Anthropomorphic thought that lies hidden in the theory of relativity or quantum physics may be so subtly concealed that it is nearly impossible to remove. The forms anthropomorphism takes in the search for extraterrestrial intelligence are much easier to detect. They are major assumptions that have become a routine part of SETI investigations. Radio astronomers search the skies for alien messages of the sort that humans might send, or humans might expect to receive. The incoming signals are supposedly broadcast using equipment of the kind that humans have devised and used. Finally, the alien messengers live in civilized societies that face problems similar to those troubling modern humans.

Searchers for signs of extraterrestrial intelligence are more susceptible to anthropomorphic thinking than other scientists. Why is this so? Their scientific training is no different from colleagues in other branches of astronomy, physics, and biology. Their subject matter, however, is radically different. Since it does not exist, it is created by the same scientists who search for proof of its existence.

Anthropomorphic thinking among SETI researchers may cause difficulties because the analogy includes an unknown, the heavens, which have been a source of speculation, and the object of human emotions, for many millennia. Comparable thinking in other arenas of the physical sciences is not so emotionally laden nor dependent upon the past.

Scientists who are confident that humans can communicate with intelligent aliens tend to overlook the difficult communication problems humans experience when confronted by *terrestrial* species. We are surrounded by organisms that share our chemistry and genes, yet we find it almost impossible to exchange information with them. There is rudimentary communication between humans and some domesticated animals—horses, cats, dogs—and scientists are working to establish a crude form of communication with apes and dolphins. If we can barely communicate with terrestrial creatures, how can we hope to decode complex messages sent by superior extraterrestrial ones?

Searchers for extraterrestrial intelligence have an answer to that question. They claim that extraterrestrial beings, unlike dogs and chimpanzees, are superior to humans and thus are able to exchange information with us. This solution simply closes the circle by bringing us back to superior beings and the religious motivation mentioned earlier.

• • •

In the fall of 2003, Dennis Overbye reported in the *New York Times* that NASA's astrobiology program placed a new emphasis on the search for extraterrestrial life. Participants in that endeavor, he noted, understood that they were held hostage by their preconceptions of life and intelligence. Overbye closed with an observation that is a fitting end to this chapter and book: "We are good at looking for things like ourselves."[5]

SOURCES OF QUOTED MATERIAL

✦

CHAPTER 1

1. Edward Grant, *Planets, Stars, and Orbs: The Medieval Cosmos, 1200–1687* (Cambridge: Cambridge University Press, 1994), 152.

2. Philip Morrison et al., eds., *The Search for Extraterrestrial Intelligence: SETI* (Washington, D.C.: NASA, 1977), vii.

3. I. S. Shklovskii and Carl Sagan, *Intelligent Life in the Universe*, trans. Paula Fern (New York: Dell, 1966), 358.

4. Richard Berenzden, ed., *Life Beyond the Earth and the Mind of Man* (Washington, D.C.: NASA, 1973), 63.

5. David W. Swift, *SETI Pioneers* (Tucson: University of Arizona Press, 1990), 57.

6. Frank D. Drake and Dava Sobel, *Is Anyone Out There? The Scientific Search for Extraterrestrial Intelligence* (New York: Delacorte, 1992), 160.

7. Ibid., 162.

8. Robert Jastrow, *The Enchanted Loom: Mind in the Universe* (New York: Simon and Schuster, 1981), 167.

9. Keay Davidson, *Carl Sagan: A Life* (New York: Wiley, 1999), 30.

10. Carl Sagan, *Planetary Exploration* (Eugene: Oregon State System of Higher Education, 1970), 67.

CHAPTER 2

1. Johann Kepler, *Kepler's Conversation with Galileo's Sidereal Messenger*, trans. E. Rosen (New York: Johnson Reprint Corporation, 1965), 43.

2. Carola Baumgardt, *Johannes Kepler: A Life and Letters* (New York: Philosophical Library, 1951), 31.

3. Johann Kepler, *Kepler's Somnium*, trans. Edward Rosen (Madison: University of Wisconsin Press, 1967), 28.

4. Baumgardt, *Kepler*, 155.

5. Scott L. Montgomery, *The Moon and the Western Imagination* (Tucson: University of Arizona Press), 188.

CHAPTER 3

1. John Wilkins, *The Discovery of a New World in the Moone* (London, 1638; repr., Delmar, NY: Scholars' Facsimiles and Reprints, 1973), 186–187.

2. Ibid., 118.

3. Paolo Rossi, "Nobility of Man and Plurality of Worlds," in *Science, Medicine and Society in the Renaissance*, ed. Allen G. Debus, vol. 2 (New York: Science History Publications, 1972), 154.

4. Bernard le Bovier de Fontenelle, *Conversations on the Plurality of the Worlds*, trans. H. A. Hargreaves (Berkeley: University of California Press, 1990), 32.

5. Ibid., 35.

6. Ibid., 35.

7. Ibid., 35.

8. Ibid., 45.

9. Christiaan Huygens, *Kosmostheoros* (The Hague, 1698). Appeared in English translation as *The Celestial Worlds Discover'd* (London, 1698; repr., London: Frank Cass & Co., 1968), 9.

10. Ibid., 61

11. Ibid., 78.

12. Walter A. McDougall, The Heavens and the Earth (New York: Basic Books, 1985), 225.

13. David E. Fisher and Marshall Jon Fisher, *Strangers in the Night: A Brief History of Life on Other Worlds* (Washington, D.C.: Counterpoint, 1998), 262.

CHAPTER 4

1. William Sheehan, *Planets & Perception: Telescopic Views and Interpretations, 1609–1909* (Tucson: University of Arizona Press, 1988), 128.

2. *Dictionary of Scientific Biography*, vol. 12, 1975, s.v. "Schiaparelli, Giovanni Virginio."

3. Michael J. Crowe, *The Extraterrestrial Life Debate, 1750–1900* (Cambridge: Cambridge University Press, 1986), 486.

4. Sheehan, *Planets & Perception*, 106.

5. William H. Pickering, "Schiaparelli's Latest Views Regarding Mars," *Astronomy and Astro-physics* 13 (Oct. 1894): 720.

6. Ibid., 721–722.
7. Crowe, *The Extraterrestrial Life Debate*, 515.
8. Ibid., 515.

CHAPTER 5

1. David Strauss, *Percival Lowell* (Cambridge, MA: Harvard University Press, 2001), 89–90.
2. Percival Lowell, *Mars* (Boston: Houghton Mifflin, 1895), 6.
3. Ibid., 209.
4. Ibid., 209.
5. Steven J. Dick, *The Biological Universe* (New York: Cambridge University Press, 1996), 93.
6. Ibid., 94.
7. Hoyt, *Lowell and Mars* (Tucson: University of Arizona Press, 1996), 297.
8. William Sheehan, *The Planet Mars* (Tucson: University of Arizona Press, 1996), 144.
9. Edward S. Morse, *Mars and Its Mystery* (Boston: Little, Brown, 1907), 143.
10. Hoyt, *Lowell and Mars*, 221.

CHAPTER 6

1. Henry S. F. Cooper, Jr., *The Search for Life on Mars* (New York: Holt, Rinehart and Winston, 1980), 40.
2. Ray Bradbury et al., *Mars and the Mind of Man* (New York: Harper & Row, 1973), 23.
3. Mark Washburn, *Mars at Last!* (New York: Putnam, 1977), 66.
4. Howard E. McCurdy, *Space and the American Imagination* (Washington, D.C.: Smithsonian Institution Press, 1997), 122.
5. Carl Sagan and Paul Fox, "The Canals of Mars: Assessment after Mariner 9," *Icarus* 25 (Aug. 1975): 602.
6. Steven J. Dick, *The Biological Universe* (New York: Cambridge University Press, 1996), 155.
7. Gerald A. Soffen, "Life on Mars?" in *The New Solar System*, ed. J. Kelly Beatty, Brian O'Leary, and Andrew Chaikin (Cambridge: Cambridge University Press, 1981) 94.
8. Cooper, *Search for Life on Mars*, 68.
9. I. S. Shklovskii and Carl Sagan, *Intelligent Life in the Universe*, trans. Paula Fern (New York: Dell, 1966), 293.
10. Ibid., 375.
11. Keay Davidson, *Carl Sagan: A Life* (New York: Wiley, 1999), 100.

12. William G. Hoyt, *Lowell and Mars* (Tucson: University of Arizona Press, 1976), 91.

13. Edgar Rice Burroughs, *A Fighting Man on Mars* (New York: Metropolitan Books, 1931), 171.

14. Davidson, *Carl Sagan*, 251.

15. Shklovskii and Sagan, *Intelligent Life in the Universe, 374.*

16. Joel Achenbach, *Captured by Aliens* (New York: Simon and Schuster, 1990), 88.

17. Dick, *The Biological Universe*, 289.

18. Christiaan Huygens, *Kosmotheoros* (The Hague, 1698). Appeared in English as *The Celestial Worlds Discover'd* (London, 1698; repr., London: Frank Cass & Co., 1968), 150–151.

19. Carl Sagan, *Cosmos* (New York: Random House, 1980), 147.

20. Davidson, *Carl Sagan,* 178.

21. Percival Lowell, *Mars* (Boston: Houghton Mifflin, 1895), 75.

CHAPTER 7

1. Ray Bradbury et al., *Mars and the Mind of Man* (New York: Harper & Row, 1973), 24.

2. Henry S. F. Cooper, Jr., *The Search for Life on Mars* (New York: Holt, Rinehart and Winston, 1980), 38.

3. Carl Sagan, *The Cosmic Connection* (New York: Dell, 1973), 133.

4. Steven J. Dick, *The Biological Universe* (New York: Cambridge University Press, 1996), 70.

5. Carl Sagan, *Other Worlds* (New York: Bantam, 1975), 79.

6. Cooper, *The Search for Life on Mars*, 27.

7. William Poundstone, *Carl Sagan: A Life in the Cosmos* (New York: Henry Holt, 1999), 198–199.

8. Bruce Murray, *Journey into Space* (New York: Norton, 1989), 69.

9. Ibid., p. 69.

10. Carl Sagan, Linda Salzman Sagan, and Frank Drake, "A Message from Earth," *Science* 175 (Feb. 25, 1972), 881.

11. Richard Berendzen, ed., *Life Beyond the Earth & the Mind of Man* (Washington, D.C.: 1973), 17.

12. John Billingham, Roger Heyns, et al., eds., *Social Implications of the Detection of an Extraterrestrial Civilization* (Mountain View, Calif.: SETI Press, 1994), 59.

CHAPTER 8

1. Steven J. Dick, *The Biological Universe* (New York: Cambridge University Press, 1996), 179.

2. E. A. Milne, *Modern Cosmology and the Christian Idea of God* (Oxford: Oxford University Press, 1952), 152–153.

3. Dick, *Biological Universe*, 348.

4. Melvin Calvin, "Communication: From Molecules tó Mars," *AIBS Bulletin*, 12 Oct. 1962, 42.

5. Simon Conway Morris, Review of *Cradle of* Life, by J. William Schopf, *New York Times Book Review*, Aug. 15, 1999, 11.

6. Dick, *Biological Universe*, 414.

7. Giuseppe Cocconi and Philip Morrison, "Searching for Interstellar Communications," *Nature* 184 (Sept. 1959): 846.

8. Graham Farmelo, ed., *It Must Be Beautiful: Great Equations of Modern Science* (London: Granta 2002), 174.

9. *History of Astronomy: An Encyclopedia*, ed. by John Lankford, s.v. "SETI," by Steven J. Dick, 458.

10. Joel Achenbach, *Captured By Aliens* (New York: Simon & Schuster, 1999), 281.

11. Christopher Chyba, "Life Beyond Mars," *Nature* 382 (Aug. 1996): 576.

12. Edward Purcell, "Radioastronomy and Communication through Space," *The Quest for Extraterrestrial Life,* ed. Donald Goldsmith (Mill Valley, Calif.: University Science Books, 1980), 196.

13. Carl Sagan, "Direct Contact among Galactic Civilizations by Relativistic Interstellar Spaceflight," in *The Quest for Extraterrestrial Life*, ed. Donald Goldsmith (Mill Valley, Calif.: University Science Books, 1980), 210.

14. Ibid., 211.

15. Freeman J. Dyson, "Search for Artificial Stellar Sources of Infrared Radiation," *Science* 131 (June 1960): 1667.

16. Ibid., 1667.

17. Freeman J. Dyson, "Search for Artificial Stellar Sources of Infrared Radiation: Addendum," in *Interstellar Communication*, ed. Alastair G. W. Cameron (New York: Benjamin, 1963), 114.

18. Freeman J. Dyson, *Disturbing the Universe* (New York: Harper & Row, 1979), 211.

19. Freeman J. Dyson, "The Search for Extraterrestrial Technology," in *Perspectives in Modern Physics*, ed. R. E. Marshak (New York: Interscience Publishers, 1966), 653.

CHAPTER 9

1. Carl Sagan, ed., *Communication with Extraterrestrial Intelligence (CETI)* (Cambridge, Mass.: MIT Press, 1973), 188.

2. Ibid., 339.

3. Ibid., 344.

4. Philip Morrison, "The Number N of Advanced Civilizations in Our Galaxy and the Question of Galactic Civilization: An Introduction," *Strategies for the Search for Life in the Universe*, ed. Michael D. Papagiannis (Dordrecht, 1980), 18.

5. Jill Tarter, "HRMS, Where We've Been, and Where We're Going," *Progress in the Search for Extraterrestrial Life*, ASP Conference Series, vol. 74. ed. G. Seth Shostak (San Francisco: Astronomical Society of the Pacific, 1995), 457.

6. John N. Wilford, "Astronomers Start Search for Life Beyond Earth," *New York Times*, Oct. 13, 1992, C2.

7. Ibid., C2.

8. John D. Boudreau, "Cosmic Endeavor or Black Hole," *Washington Post*, Feb. 15, 1994, E, 1.

9. John N. Wilford, "Ear to the Universe Is Plugged by Budget Cutters," *New York Times*, Oct. 7, 1993, 12.

CHAPTER 10

1. David Hume, *Dialogues Concerning Natural Religion and the Natural History of Religion*, ed. J. C. A. Gaskin (Oxford: Oxford University Press, 1993), 141.

2. Edward Purcell, "Radioastronomy and Communication through Space," *The Quest for Extraterrestrial Life*, ed. Donald Goldsmith (Mill Valley, Calif.: University Science Books, 1980), 195.

3. Sheldon Lee Glashow, "The Death of Science!?," *The End of Science?*, ed. Richard Q. Elvee (Lanham, Md.: University Press of America, 1992), 28.

4. Steven Weinberg, "Illustrative Prediction," *Science* 285 (Aug. 1999): 1013.

5. Barry Allen, "What It's All About," *Science* 285 (July 1999): 205.

6. Steven Pinker, *How the Mind Works* (New York: Norton, 1997), 153.

7. Robert Bieri, "Huminoids on Other Planets," *American Scientist* 52 (Dec. 1964): 457.

8. Simon Conway Morris, *Life's Solution: Inevitable Humans in a Lonely Universe* (Cambridge: Cambridge University Press, 2003), 32.

9. John A. Ball, "The Zoo Hypothesis," *Icarus* 19 (Mar. 1973), 348.

10. I. S. Shklovskii and Carl Sagan, *Intelligent Life in the Universe*, trans. Paula Fern (New York: Dell, 1966), 375.

CHAPTER 11

1. Carl Sagan, "Unidentified Flying Object," *The Encyclopedia Americana* (1982), vol. 27, 369.

2. Cynthia Ozick, "If You Can Read This, You Are Too Far Out," *Esquire* 79 (Jan. 1973), 74.

3. David W. Swift, *SETI Pioneers* (Tucson: University of Arizona Press, 1990) 83.

4. Joel Achenbach, *Captured by Aliens* (New York: Simon & Schuster, 1999), 283.

5. Dennis Overbye, "Where Are Those Aliens?" *New York Times*, Nov. 11, 2003, F15.

BIBLIOGRAPHY

✦

Bibliographic Note: The items named hereafter record my debt to those who have written on the topics I discuss. The names of the authors are not arranged alphabetically. They are listed according to the use I make of their research within a given chapter.

At this point I wish to acknowledge the consistent use I have made of an excellent reference work: David Darling, *The Extraterrestrial Encyclopedia: An Alphabetic Reference to All Life in the Universe* (New York: Three Rivers Press, 2000). Based on 634 books, articles, and conference papers, this volume provides a comprehensive guide to its subject.

CHAPTER I

The Ideas

Steven J. Dick, *Plurality of Worlds: The Origins of the Extraterrestrial Debate from Democritus to Kant* (Cambridge: Cambridge University Press, 1982.); Michael J. Crowe, *The Extraterrestrial Life Debate, 1759–1900* (Cambridge: Cambridge University Press, 1986); Karl S. Guthke, *The Last Frontier*, trans. Helen Atkins (Ithaca, N.Y.: Cornell University Press, 1990).

The Infinitization of the Universe

Milton K. Munitz, "One Universe or Many?" *Journal of the History of Ideas* 12 (Apr. 1951): 231–255; Cyril Bailey, *The Greek Atomists and Epicurus* (Oxford: Oxford University Press, 1928); David C. Lindberg, *The Beginnings of Western Science* (Chicago: University of Chicago Press, 1992); Pierre Duhem, *Medieval Cosmology*, ed. and trans. Roger Ariew (Chicago: University of Chicago Press, 1985); Edward Grant, *Much Ado about Nothing: Theories of Space and Vacuum from the Middle Ages to the Scientific Revolution* (Cambridge: Cambridge University Press, 1981), *Planets, Stars, and Orbs: The Medieval Cosmos, 1200–1687* (Cambridge: Cambridge University Press, 1994), and *The Foundations of Modern*

Science in the Middle Ages (New York: Cambridge University Press, 1996); Edward Grant, ed., *A Source Book in Medieval Science* (Cambridge, Mass.: Harvard University Press, 1974); Alexandre Koyré, *From the Closed World to the Infinite Universe* (Baltimore: Johns Hopkins Press, 1957); Thomas S. Kuhn, *The Copernican Revolution* (Cambridge, Mass.: Harvard University Press, 1957).

Other Worlds, Other Life

Marjorie Hope Nicolson, *Voyages to the Moon* (New York: Macmillan, 1960); Plutarch, *Moralia*, XII, trans. Harold Cherniss (Cambridge, Mass.: Harvard University Press, 1927); Samuel Sambursky, *The Physical World of the Greeks* (London: Routledge and Paul, 1956); Donald A. Russell, *Plutarch* (New York: Scribner, 1973); Edward Grant, *Planets, Stars, and Orbs: The Medieval Cosmos, 1200–1687* (Cambridge: Cambridge University Press, 1994); Edward Grant, ed., *A Source Book in Medieval Science* (Cambridge, Mass.: Harvard University Press, 1974); C. S. Lewis, *The Discarded Image* (Cambridge: Cambridge University Press, 1964); Philip Morrison et al., eds., *The Search for Extraterrestrial Intelligence, SETI* (Washington, D.C.: NASA, 1977); Grant McCooley and H. W. Miller, "Saint Bonaventure, Francis Mayron, William Vorilong, and the Doctrine of a Plurality of Worlds," *Speculum* 12 (July 1937): 386–389; David E. Fisher and Marshall Jon Fisher, *Strangers in the Night: A Brief History of Life on Other Worlds* (Washington, D.C.: Counterpoint, 1998); Edward Grant, "Aristotelianism and the Longevity of the Medieval World View," *History of Science* 16 (June 1978): 93–106; Cardinal Nicolas of Cusa, *Of Learned Ignorance*, trans. Germain Heron (London: Routledge and Paul, 1954); Francis R. Johnson, *Astronomical Thought in Renaissance England* (Baltimore: Johns Hopkins Press, 1937); Thomas S. Kuhn, *The Copernican Revolution* (Cambridge, Mass.: Harvard University Press, 1957); Dorothy Waley Singer, *Giordano Bruno: His Life and Thought* (New York: Schuman, 1950); Karsten Harries, *Infinity and Perspective* (Cambridge, Mass.: MIT Press, 2001); Gary B. Ferguson, *Science and Religion: A Historical Introduction* (Baltimore: Johns Hopkins University Press, 2000).

The Principle of Mediocrity

I. S. Shklovskii and Carl Sagan, *Intelligent Life in the Universe*, trans. Paula Fern (New York: Dell, 1966).

Superior Beings

Richard Berendzen, ed., *Life beyond the Earth and the Mind of Man* (Washington, D.C.: NASA, 1973); Edward Grant, *Planets, Stars, and Orbs: The Medieval Cosmos, 1200–1687* (Cambridge: Cambridge University Press, 1994); Edward Grant, ed., *A Source Book in Medieval Science* (Cambridge, Mass.: Harvard University Press, 1974); Thomas S. Kuhn, *The Copernican Revolution* (Cambridge, Mass.: Harvard University Press, 1957.

The Religious Impulse

David W. Swift, ed. *SETI Pioneers* (Tucson: University of Arizona Press, 1990); Frank Drake and Dava Sobel, *Is Anyone Out There?* (New York: Delacorte, 1992); Marina

Benjamin, *Rocket Dreams* (New York: Free Press, 2003); Robert Jastrow, *The Enchanted Loom* (New York: Simon & Schuster, 1981); Keay Davidson, *Carl Sagan: A Life* (New York: Wiley, 1999); Carl Sagan, *Planetary Exploration* (Eugene: Oregon State System of Higher Education, 1970); Robert Plank, *The Emotional Significance of Imaginary Beings* (Springfield: Charles C. Thomas, 1968).

CHAPTER 2

New Ways of Seeing, Old Ways of Thinking

Henry C. King, *The History of the Telescope* (Cambridge, Mass.: Sky Publication Corporation, 1955).

Galileo's Telescope

Galileo Galilei, *Sidereus Nuncius, or the Sidereal Messenger*, trans. Albert van Helden (Chicago: Chicago University Press, 1989).

Kepler and Galileo

Karl S. Guthke, *The Last Frontier*, trans. Helen Atkins (Ithaca, N.Y.: Cornell University Press, 1990); Johannes Kepler, *Kepler's Conversation with Galileo's Sidereal Messenger*, trans. Edward Rosen (New York: Johnson Reprint Corporation, 1965); Steven J. Dick, *Plurality of Worlds: The Origins of the Extraterrestrial Life Debate from Democritus to Kant* (Cambridge: Cambridge University Press, 1982); Stillman Drake, *Galileo at Work: His Scientific Biography* (Chicago: Chicago University Press, 1978); Gerald Holton, "Johannes Kepler's Universe: Its Physics and Metaphysics," *American Journal of Physics* 24 (May 1956): 340–351.

Kepler's Dream

Johannes Kepler, *Kepler's Somnium*, trans. Edward Rosen (Madison: University of Wisconsin Press, 1967); Marjorie Hope Nicolson, *Voyages to the Moon* (New York: Macmillan, 1960); Carola Baumgardt, *Johannes Kepler: His Life and Letters* (New York: Philosophical Library, 1951).

Science Fact and Science Fiction

Max Caspar, *Kepler*, trans. C. Doris Hellman (London: Abelard-Schuman, 1959); Fernand Hallyn, *The Poetic Structure of the World: Copernicus and Kepler* (Cambridge, Mass.: MIT Press, 1990); Christopher Duffy, *Siege Warfare: The Fortress in the Early Modern World, 1494–1600* (London: Routledge & Kegan Paul, 1979); Jim Bennett and Stephen Johnston, *The Geometry of War, 1500–1750* (Oxford: Museum of the History of Science, 1996); A. W. Richeson, *English Land Measuring to 1800* (Cambridge, Mass.: MIT Press, 1966).

Mapping the Moon

Scott L. Montgomery, *The Scientific Voice* (New York: Guilford Press, 1996); Norman J. W. Thrower, *Maps and Civilization* (Chicago: University of Chicago Press, 1996);

Z. Kopal and R. W. Carder, *Mapping of the Moon* (Dordrecht: Reidel, 1974); Ivan Volkoff, *Johnannes Hevelius and His Catalog of Stars* (Provo: Brigham University Press, 1971); Scott L. Montgomery, *The Moon and the Western Imagination* (Tucson: University of Arizona Press, 1999).

CHAPTER 3

Bishop Wilkins's Voyage to the Moon

John Wilkins, *The Discovery of a World in the Moone*, 1638 (reprint, Delmar, N.Y.: Scholars' Facsimiles & Reprints, 1973); Barbara J. Shapiro, *John Wilkins, 1614–1672* (Berkeley: University of California Press, 1969), and *Probability and Certainty in Seventeenth-Century England* (Princeton: Princeton University Press, 1983); Marjorie Hope Nicolson, *Voyages to the Moon* (New York: Macmillan, 1960).

Descartes' Cosmic Model

Steven J. Dick, *Plurality of Worlds: The Origins of the Extraterrestrial Debate from Democritus to Kant* (Cambridge: Cambridge University Press, 1982); Karl S. Guthke, *The Last Frontier*, trans. Helen Atkins (Ithaca, N.Y.: Cornell University Press, 1990); Michael J. Crowe, *The Extraterrestrial Life Debate: 1750–1900* (Cambridge: Cambridge University Press, 1986).

The Plurality of the Worlds: Fontenelle

M. de Fontenelle (Bernard le Bovier), *Conversations on the Plurality of Worlds*, trans. H. A. Hargreaves (Berkeley: University of California Press, 1990); Steven J. Dick, *The Biological Universe* (New York: Cambridge University Press, 1996); Franklin Thomas, *The Environmental Basis of Society* (New York: Century, 1925).

Huygens's "Probable Conjectures"

Henry Oldenburg, *The Correspondence of Henry Oldenburg*, vol. 1, ed. A. R. Hall and M. B. Hall (Madison: University of Wisconsin Press, 1965); Lorraine Daston, *Classical Probability in the Enlightenment* (Princeton: Princeton University Press, 1988); Christiaan Huygens, *Kosmotheoros* (The Hague, 1698), appeared in English translation as *The Celestial Worlds Discover'd* (London, 1698; reprint, London: Frank Cass & Co., 1968); Barbara J. Shapiro, *Probability and Certainty in Seventeenth-Century England* (Princeton: Princeton University Press, 1983); H. J. M. Bos et al., eds., *Studies on Christiaan Huygens* (Lisse: Swets & Zertlinger, 1980); Valentin Boss, *Newton and Russia* (Cambridge, Mass.: Harvard University Press, 1972); Svetlana Alpers, *The Art of Describing: Dutch Art in the Seventeenth Century* (Chicago: Chicago University Press, 1983); Carl Sagan, *Cosmos* (New York: Random House, 1980); Simon Schama, *The Embarrassment of Riches: An Interpretation of Dutch Culture in the Golden Age* (New York: Knopf, 1987).

The Maritime Analogy

Werhner von Braun, *The Mars Project* (Urbana: University of Illinois Press, 1953); William E. Burrows, *Exploring Space: Voyages in the Solar System and Beyond* (New York:

Random House, 1990); Erlend A. Kennan and Edmund H. Harvey, Jr., *Mission to the Moon* (New York: Morrow, 1969); Lloyd S. Swenson, Jr., et al., eds., *This New Ocean: A History of Project Mercury* (Washington, D.C.: NASA, 1966); Richard S. Lewis, *The Voyages of Apollo* (New York: Quadrangle, 1974); Walter A. McDougall, *The Heavens and the Earth* (New York: Basic Books, 1985); Robert Plank, *The Emotional Significance of Imaginary Beings* (Springfield, Ill.: Charles C. Thomas, 1968); Valerie Neal, ed., *Where Next Columbus? The Future of Space Exploration* (New York: Oxford University Press, 1994); Carl Sagan, *Pale Blue Dot: A Vision of the Human Future in Space* (New York: Random House, 1994).

CHAPTER 4

The Development of Planetary Astronomy

Michael J. Crowe, *The Extraterrestrial Life Debate, 1750–1900* (Cambridge: Cambridge University Press, 1986); Karl S. Guthke, *The Last Frontier*, trans. Helen Atkins (Ithaca, N.Y.: Cornell University Press, 1990); Simon Schaffer, "The Great Laboratories of the Universe: William Herschel on Matter, Theory, and Planetary Life," *Journal for the History of Astronomy* 11 (June 1980): 81–111; William Sheehan, *The Planet Mars: A History of Observation and Discovery* (Tucson: University of Arizona Press, 1996); Asaph Hall, "The Discovery of the Satellites of Mars," *Monthly Notices of the Royal Astronomical Society* 38 (Feb. 1878): 205–209.

Through the Eyepiece

William Sheehan, *Planets and Perceptions: Telescopic Views and Interpretations, 1609–1909* (Tucson: University of Arizona Press, 1988).

Schiaparelli's Canals

Michael J. Crowe, The *Extraterrestrial Life Debate, 1750–1900* (Cambridge: Cambridge University Press, 1986); Giorgio Abetti, "Schiaparelli, Giovanni Virginio," in *Dictionary of Scientific Biography*, vol. 12, ed. Charles C. Gillispie, 161 (New York: C. Scribners & Sons, 1970–1990); William Sheehan, *Planets and Perception: Telescopic Views and Interpretations, 1609–1909* (Tucson: University of Arizona Press, 1988); John Pudney, *Suez: De Lesseps' Canal* (New York: Prager, 1969); David G. McCullough, *The Path between the Seas: The Creation of the Panama Canal, 1870–1914* (New York: Simon & Schuster, 1977); Charles Hadfield, *World Canals: Inland Navigation Past and Present* (Newton Abbot, U.K.: David & Charles, 1986); William Sheehan, *The Planet Mars: A History of Observation and Discovery* (Tucson: University of Arizona Press, 1996); William Graves Hoyt, *Lowell and Mars* (Tucson: University of Arizona Press, 1976); Jane Kate Leonard, *Controlling from Afar: The Daoguang Emperor's Management of the Grand Canal Crisis, 1824–1826* (Ann Arbor: University of Michigan Press, 1996); Steven J. Dick, *The Biological Universe* (New York: Cambridge University Press, 1996).

The Martian Engineers

Michael J. Crowe, The *Extraterrestrial Life Debate, 1750–1900* (Cambridge: Cambridge University Press, 1986); William H. Pickering, "Schiaparelli's Latest Views Regarding Mars," *Astronomy and Astro-physics* 13 (Oct. 1894): 632–640, 714–723; Giovanni V. Schiaparelli, "La vie sur la planète Mars," *Société astronomique de France* 12 (1898): 423–429.

CHAPTER 5

The Orientalist

William Graves Hoyt, *Lowell and Mars* (Tucson: University of Arizona Press, 1976); Norriss S. Hetherington, "Percival Lowell: Professional Scientist or Interloper?" *Journal of the History of Ideas* 42 (Jan.–Mar. 1981): 159–161; John Lankford, "Education of Astronomers," in John Lankford, ed., *History of Astronomy: An Encyclopedia* (New York: Garland, 1997); David Strauss, " 'The Far East' in the American Mind, 1883–1894: Percival Lowell's Decisive Impact," *Journal of American-East Asian Relations* 2 (fall 1993): 217–241; Gerald A. Figal, *Percival Lowell's Analysis of the Japanese Soul* (senior thesis, University of California, Santa Barbara, 1984); Robert S. Ellwood, "Percival Lowell's Journey to the East," *Sewanee Review* 78 (spring 1970): 285–309; Percival Lowell, *The Soul of the Far East* (Boston: Houghton Mifflin, 1888); A. Lawrence Lowell, *Biography of Percival Lowell* (New York: Macmillan, 1935); Ferris Greenslet, *The Lowells and Their Seven Worlds* (Boston: Houghton Mifflin, 1946); David Strauss, *Percival Lowell* (Cambridge, Mass.: Harvard University Press, 2001).

The Creation of Mars

Michael J. Crowe, *The Extraterrestrial Life Debate, 1750–1900* (Cambridge: Cambridge University Press, 1986); William Sheehan, *The Planet Mars: A History of Observation and Discovery* (Tucson: University of Arizona Press, 1996), and *Planets and Perception: Telescopic Views and Interpretations, 1609–1909* (Tucson: University of Arizona Press, 1988); Edward S. Holden, "The Lowell Observatory in Arizona," *Astronomical Society of the Pacific Publications* 36 (1894): 160–169; William Lowell Putnam, *The Explorers of Mars Hill: A Centennial History of Lowell Observatory* (West Kennebunk, Maine: 1994); David Strauss, "Percival Lowell, W. H. Pickering and the Founding of Lowell Observatory," *Annals of Science* 51 (Jan. 1994): 37–58; A. Lawrence Lowell, *Biography of Percival Lowell* (New York: Macmillan, 1935); Ferris Greenslet, *The Lowells and Their Seven Worlds* (Boston: Houghton Mifflin, 1946); Steven J. Dick, *The Biological Universe* (New York: Cambridge University Press, 1996); William Graves Hoyt, *Lowell and Mars* (Tucson: University of Arizona Press, 1976); William C. Heffernan, "The Singularity of Our Inhabited World: William Whewell and A. R. Wallace in Dissent," *Journal of the History of Ideas* 39 (Jan.–Mar. 1978): 81–100; Percival Lowell, *Mars* (Boston: Houghton, Mifflin and Co., 1895), *Mars and Its Canals* (New York: Macmillan, 1906), *Mars as the Abode of Life* (New York: Macmillan, 1910), and *The Evolution of Worlds* (New York: Macmillan, 1909); Karl S.

Guthke, *The Last Frontier*, trans. Helen Atkins (Ithaca, N.Y.: Cornell University Press, 1990); David Strauss, *Percival Lowell* (Cambridge, Mass.: Harvard University Press, 2001); Norriss S. Hetherington, "Lowell's Theory of Life on Mars," *Astronomical Society of the Pacific*, leaflet no. 501, Mar. 1971.

The Eye of the Beholder

Michael J. Crowe, *The Extraterrestrial Life Debate, 1750–1900* (Cambridge: Cambridge University Press, 1986); William Graves Hoyt, *Lowell and Mars* (Tucson: University of Arizona Press, 1976); William Sheehan, *Planets and Perception: Telescopic Views and Interpretations, 1609–1909* (Tucson: University of Arizona Press, 1988), and *The Planet Mars: A History of Observation and Discovery* (Tucson: University of Arizona Press, 1996); E. M. Antoniadi, *La Planète Mars* (Paris: Hermann éditeurs des sciences et des arts, 1930).

Lowell's America

George Ernest Webb, "The Planet Mars and Science in Victorian America," *Journal of American Culture* 3 (winter 1980): 573–580; Hillel Schwartz, *Century's End: A Cultural History of the Fin de Siècle* (New York: Doubleday, 1990); T. J. Jackson Lears, *No Place of Grace* (New York: Pantheon, 1981); John F. Kasson, *Civilizing the Machine* (New York: Grossman Publishers, 1976); William H. Jordy, *Henry Adams: Scientific Historian* (New Haven: Yale University Press, 1952); I. F. Clarke, *Voices Prophesying War: Future Wars 1763–3749* (New York: Oxford University Press, 1992); Camille Flammarion, *Omega: The Last Days of the World*, 1894, translation of *La fin du monde* (New York: Arno Press, 1975); Clinton Hart Merriam, *Selected Works of Clinton Hart Merriam* (New York: Arno Press, 1974); Walter Prescott Webb, *The Great Plains* (Boston: Ginn and Co., 1931); William E. Warne, *The Bureau of Reclamation* (New York: Prager, 1973); William E. Smythe, *The Conquest of Arid America* (New York: Macmillan & Co., 1911); Charles Hadfield, *World Canals: Inland Navigation Past and Present* (Newton Abbot, U.K.: David & Charles, 1986); Lester F. Ward, "Mars and Its Lesson," *Brown Alumni Monthly* 7 (Mar. 1907): 159–165, and *Applied Sociology* (Boston: Ginn & Co., 1906); anonymous, "Why the Dwellers on Mars Do Not Make War," *Current Literature* 42 (Feb. 1907): 211–214; Richard Hofstadter, *Social Darwinism in American Thought* (New York: George Braziller, 1959); Ferris Greenslet, *The Lowells and Their Seven Worlds* (Boston: Houghton Mifflin, 1946); David Strauss, "Fireflies Flashing in Unison; Percival Lowell, Edward Morse and the Birth of Planetology," *Journal for the History of Astronomy* 24 (Aug. 1993): 158–169; Edward S. Morse, *Mars and Its Mystery* (Boston: Little, Brown, 1907).

CHAPTER 6

Lowell's Legacy

David Strauss, *Percival Lowell* (Cambridge, Mass.: Harvard University Press, 2001); Ferris Greenslet, *The Lowells and Their Seven Worlds* (Boston: Houghton Mifflin, 1946); William

Graves Hoyt, *Lowell and Mars* (Tucson: University of Arizona Press, 1976); Michael J. Crowe, *The Extraterrestrial Life Debate, 1750–1900* (Cambridge: Cambridge University Press, 1986); Karl S. Guthke, *The Last Frontier,* trans. Helen Atkins (Ithaca, N.Y.: Cornell University Press, 1990); Steven J. Dick, *The Biological Universe* (New York: Cambridge University Press, 1996); Keay Davidson, *Carl Sagan: A Life* (New York: Wiley, 1999); H. G. Wells, *The War of the Worlds*, ed. David Y. Hughes (New York: Oxford University Press, 1995); Kurd Lasswitz, *Two Planets, Auf Zwei Planeten*, trans. Hans H. Rudnick (Carbondale: Southern Illinois Press, 1971); Mark R. Hillegas, "Martians and Mythmakers: 1877–1938," in *Challenges in American Culture*, ed. Ray B. Browne, et al. (Bowling Green, Ohio: Bowling Green University Popular Press, 1970); Franz Rottensteiner, "Kurd Lasswitz: A German Pioneer of Science Fiction," in Thomas D. Clareson, ed., *SF: The Other Side of Realism* (Bowling Green, Ohio: Bowling Green University Popular Press, 1971); Richard D. Mullen, "The Undisciplined Imagination: Edgar Rice Burroughs and Lowellian Mars," in Thomas D. Clareson ed., *SF: The Other Side of Realism* (Bowling Green, Ohio: Bowling Green University Popular Press, 1971); Wiley Ley, *Rockets, Missiles, and Space Travel* (New York: Viking Press, 1951); David W. Swift, *SETI Pioneers* (Tucson: University of Arizona Press, 1990); Henry S. F. Cooper, Jr., *The Search for Life on Mars* (New York: Holt, Rinehart and Winston, 1980); Hubertus Strughold, *The Green and Red Planet* (Albuquerque: University of New Mexico Press, 1953); Ray Bradbury, *Mars and the Mind of Man* (New York: Harper & Row, 1973); David E. Fisher and Marshall Jon Fisher, *Strangers in the Night: A Brief History of Life on Other Worlds* (Washington, D.C.: Counterpoint, 1998); Howard E. McCurdy, *Space and the American Imagination* (Washington, D.C.: Smithsonian Institution Press, 1997); Edward C. Ezell and Linda N. Ezell, *On Mars: Exploration of the Red Planet, 1958–1978* (Washington, D.C.: NASA, 1984); William Sheehan, *The Planet Mars: A History of Observation and Discovery* (Tucson: University of Arizona Press, 1996); Carl Sagan, "Mars—A New World to Explore," *National Geographic* 132 (Dec. 1967): 821–841.

Hopes Dashed

Carl Sagan and Paul Fox, "The Canals of Mars: Assessment after Mariner 9," *Icarus* 25 (Aug. 1975): 602–612; Norman H. Horowitz, *To Utopia and Back: The Search for Life in the Solar System* (New York: W. H. Freeman, 1989); Bruce Murray, *Journey into Space: The First Three Decades of Space Exploration* (New York: Norton, 1989); Monica Grady, Ian Wright, and Colin Pillinger, "Opening a Martian Can of Worms?" *Nature* 382 (Aug. 15, 1996): 575–576; Richard A. Kerr, "Ancient Life on Mars?" Science 273 (Aug. 16, 1996): 864–866; Donald Goldsmith, *The Hunt for Life on Mars* (New York: Dutton, 1997); J. William Schopf, *Cradle of Life: The Discovery of the Earth's Earliest Fossils* (Princeton: Princeton University Press, 1999); John Noble Wilford, "Spacecraft Sends Its First Images of Mars," *New York Times* (Mar. 2, 2002): A, 12; Richard A. Kerr, "A Tale of Two Landings, One Orbiting," *Science* 303 (Jan. 9, 2004): 150–151; Andrew Lawler, "Scientists Add Up Gains and Losses in Bush's New Vision for NASA," *Science* 303 (Jan. 23, 2004): 444–445; Linda Rowan, "Opportunity Runneth Over," *Science* 306 (Dec. 3, 2004): 1697; Kenneth Chang,

"Methane in Martian Air Suggests Life Beneath the Surface," *New York Times* (Nov. 23, 2004): Science Times, 2.

Lowell's Successor: Carl Sagan

David Strauss, *Percival Lowell* (Cambridge, Mass.: Harvard University Press, 2001); Henry S. F. Cooper, Jr., *The Search for Life on Mars* (New York: Holt, Rinehart and Winston, 1980); Carl Sagan, "The Planet Venus," *Science* 133 (Jan. 23, 1961): 849–858; Yerzant Terizan and Elizabeth M. Bilson, eds., *Carl Sagan's Universe* (Cambridge: Cambridge University Press, 1997); I. S. Shklovskii and Carl Sagan, *Intelligent Life in the Universe*, trans. Paula Fern (New York: Dell, 1966); Keay Davidson, *Carl Sagan: A Life* (New York: Wiley, 1999); Carl Sagan, Jonathan Norton Leonard, and the editors of Time-Life Books, *Planets* (New York: Time-Life Books, 1969); Howard E. McCurdy, *Space and the American Imagination* (Washington, D.C.: Smithsonian University Press, 1997); Steven J. Dick, *The Biological Universe* (New York: Cambridge University Press, 1996); Edward Clinton Ezell and Linda Neuman Ezell, *On Mars: Exploration of the Red Planet, 1958–1978* (Washington, D.C.: NASA, 1984); Carl Sagan, *The Cosmic Connection: An Extraterrestrial Perspective* (New York: Doubleday, 1973), and *Other Worlds* (New York: Bantam, 1975); William Sheehan, *The Planet Mars: A History of Observation and Discovery* (Tucson: University of Arizona Press, 1996); Walter A. McDougall, *The Heavens and the Earth: A Political History of the Space Age* (New York: Basic Books, 1985); William Graves Hoyt, *Lowell and Mars* (Tucson: University of Arizona Press, 1976); Hughes H. Kieffer, Bruce M. Jakosky, Conway Snyder, and Mildred S. Matthews, eds., *Mars* (Tucson: University of Arizona Press, 1992); Peter John Cattermole, *Mars: The Story of the Red Planet* (London: Chapman and Hall, 1992); Edgar Rice Burroughs, *A Fighting Man of Mars* (New York: Metropolitan Books, 1931).

The Rational Speculator

Carl Sagn, *The Demon-Haunted World* (New York: Random House, 1996); Joel Achenbach, *Captured by Aliens* (New York: Simon & Schuster, 1999); Philip Morrison, "The Taste for Speculation," *Science* 153 (Sept. 30, 1966): 1628–1629; Christiaan Huygens, *Kosmotheoros* (The Hague 1698), appeared in English translation as *The Celestial Worlds discover'd* (London: Cass, 1968); Percival Lowell, *Mars* (Boston: Houghton Mifflin, 1895); Donald Goldsmith, *The Hunt for Life on Mars* (New York: Dutton, 1997).

CHAPTER 7

Exobiology

Steven J. Dick and James E. Strick, *The Living Universe: NASA and the Development of Astrobiology* (New Brunswick, N.J.: Rutgers University Press, 2004); Keay Davidson, *Carl Sagan: A Life* (New York: Wiley, 1999); Joshua Lederberg, "Exobiology: Approaches to Life beyond the Earth," *Science* 132 (Aug. 12, 1960): 393–400; George Gaylord Simpson, "The Nonprevalance of Humanoids," *Science* 143 (Feb. 21, 1964): 769–775; Roger D.

Launius, "A Western Mormon in Washington, D.C.: James Fletcher, NASA and the Final Frontier," *Pacific Historical Review* 64 (May 1995): 217–241; Erich Robert Paul, *Science, Religion and Mormon Cosmology* (Urbana: University of Illinois Press, 1992); Carl Sagan, "Biological Contamination of the Moon," *Proceedings of the National Academy of Sciences* 46 (Apr. 15, 1960): 396–402; Edgar M. Cortright, ed., *Apollo Expeditions to the Moon* (Washington, D.C.: NASA, 1975); Carl Sagan and Joshua Lederberg, "The Prospects for Life on Mars: A Pre-Viking Assessment," *Icarus* 28 (June 1976): 291–300; Colin S. Pittendrigh, Wolf Vishniac, and J. P. T. Pearman, eds., *Biology and the Exploration of Mars* (Washington, D.C.: National Academy of Sciences, 1966); Bruce Murray, *Journey into Space* (New York: Norton, 1989); Edward C. Ezell and Linda N. Ezell, *On Mars: Exploration of the Red Planet, 1958–1978* (Washington, D.C.: NASA, 1984); William K. Hartmann and Odell Raper, *The New Mars: The Discoveries of Mariner 9* (Washington, D.C.: NASA, 1974); Henry S. F. Cooper, Jr., *The Search for Life on Mars* (New York: Holt, Rinehart and Winston, 1980); Norman H. Horowitz, *To Utopia and Back: The Search for Life in the Solar System* (New York: W. H. Freeman, 1986); Steven J. Dick, *The Biological Universe* (Cambridge: Cambridge University Press, 1996); Ray Bradbury, *Martian Chronicles* (New York: Doubleday, 1950); Carl Sagan, *Cosmic Connection: An Extraterrestrial Perspective* (New York: Doubleday, 1973), and *Other Worlds* (New York: Bantam, 1975); anonymous, "Mars: The Search Begins," *Time* 108 (July 5, 1976): 87–90; Clark A. Brady, *The Burroughs Encyclopedia* (Jefferson, N.C.: McFarland, 1996); Michael H. Hart, and Ben Zuckerman, eds., *Extraterrestrials: Where Are They?* (New York: Pergamon Press, 1982).

The Showman

Rae Goodell, *The Visible Scientists* (Boston: Little, Brown, 1977); Carl Sagan, *The Cosmic Connection: An Extraterrestrial Perspective* (New York: Doubleday, 1973); Steven J. Dick, *The Biological Universe* (New York: Cambridge University Press, 1996); Carl Sagan, *Other Worlds* (New York: Bantam, 1975); Henry S. F. Cooper, Jr., *The Search for Life on Mars* (New York: Holt, Rinehart and Winston, 1980); William Poundstone, *Carl Sagan: A Life in the Cosmos* (New York: Henry Holt, 1999); Bruce Murray, *Journey into Space* (New York: Norton, 1989).

Message to the Stars

Keay Davidson, *Carl Sagan: A Life* (New York: Wiley, 1999); David Morrison and Jane Samz, *Voyages to Jupiter* (Washington, D.C.: NASA, 1980); Richard O. Fimmel, James van Allen, and Eric Burgess, *Pioneer: First to Jupiter, Saturn, and Beyond* (Washington, D.C.: NASA, 1980); Carl Sagan, Linda Salzman Sagan, and Frank D. Drake, "A Message from Earth," *Science* 175 (Feb. 25, 1972): 881–884; Carl Sagan, *Cosmic Connection* (New York: Doubleday, 1973), and *Murmurs of Earth: The Voyager Interstellar Record* (New York: Random House, 1978); E. H. Gombrich, "The Visual Image," *Scientific American* 227 (Sept. 1972): 82–96; David Lamb, *The Search for Extraterrestrial Intelligence: A Philosophical Inquiry* (London: Routledge, 2001); Marina Benjamin, *Rocket Dreams* (New York: Free Press, 2003).

After Contact

Donald N. Michael, *Proposed Studies on the Implications of Peaceful Space Activities for Human Affairs*, prepared for the National Aeronautics and Space Administration by the Brookings Institute, Report of the Committee on Science and Aeronautics, U.S. House of Representatives, 87th Congress, 1st Session, March 24, 1961 (Washington, D.C., 1961); Alastair G. W. Cameron, ed., *Interstellar Communication* (New York: W. A. Benjamin, 1963); Carl Sagan, ed., *Communication with Extraterrestrial Intelligence (CETI)* (Cambridge, Mass.: MIT Press, 1973); Carl Sagan, *Broca's Brain: Reflections on the Romance of Science* (New York: Random House, 1979); Keay Davidson, *Carl Sagan: A Life* (New York: Wiley, 1999); Richard Berendzen, ed., *Life beyond Earth and the Mind of Man* (Washington, D.C.: NASA, 1973); Bernard Oliver, and John Billingham, *Project Cyclops* (Stanford: Stanford University, 1972); Carl Sagan, *Cosmos* (New York: Random House, 1980), and "Extraterrestrial Intelligence: An International Petition," *Science* 218 (Oct. 29, 1982): 426; Edward Regis, Jr., "SETI Debunked," in Edward Regis, Jr., ed., *Extraterrestrials: Science and Alien Intelligence* (Cambridge: Cambridge University Press, 1985); Carl Sagan, "The Search for Who We Are," *Discover* 3 (Mar. 1982): 31–33, and "We Are Nothing Special," *Discover* 4 (Mar. 1983): 30–36; Carl Sagan and William I. Newman, "The Solipsist Approach to Extraterrestrial Intelligence," *Quarterly Journal of the Royal Astronomical Society* 24 (June 1983): 113–121.

The End of the Beginning

John Billingham, Roger Heyns, et al., eds., *Social Implications of the Detection of an Extraterrestrial Civilization* (Mountain View, Calif.: SETI Press, 1999); G. Seth Shostak, ed., *Progress in the Search for Extraterrestrial Life, Astronomical Society of the Pacific Conference Series* 74 (San Francisco, 1995), and *Sharing the Universe* (Berkeley: Berkeley Hills Books, 1998).

CHAPTER 8

Skepticism and Acceptance

Steven J. Dick, *The Biological Universe* (Cambridge: Cambridge University Press, 1996); Steven J. Dick and James E. Strick, *The Living Universe: NASA and the Development of Astrobiology* (New Brunswick, N.J.: Rutgers University Press, 2004); Frank J. Tipler, "A Brief History of the Extraterrestrial Intelligence Concept," *Quarterly Journal of the Royal Astronomical Society* 22 (Jun. 1981): 133–145; H. Spencer Jones, *Life on Other Worlds* (New York: Macmillan, 1940); E. A. Milne, *Modern Cosmology and the Christian Idea of God* (Oxford: Oxford University Press, 1952); William Graves Hoyt, *Lowell and Mars* (Tucson: University of Arizona Press, 1976); Robert W. Smith, *The Expanding Universe* (Cambridge: Cambridge University Press, 1982), and "Edwin P. Hubble and the Transformation of Cosmology," *Physics Today* 43 (Apr. 1990): 52–58.

Origins of Life

Melvin Calvin, *Chemical Evolution* (Oxford: Oxford University Press, 1969), "Chemical Evolution and the Origin of Life," *American Scientist* 44 (July 1956): 248–263, "Round Trip from Space," *Evolution* 13 (Sept. 1959): 362–377, "Communication: From Molecules to Mars," *AIBS Bulletin* 12 (Oct. 1962): 29–44, "Talking to Life on Other Worlds," *Science Digest* 53 (Jan. 1963): 14–19, 88, and "What a Man from Space Looks Like," *Science Digest* 53 (Feb. 1963): 88–89; Joel Achenbach, *Captured by Aliens* (New York: Simon & Schuster, 1999); Eric J. Chaisson, *Cosmic Evolution: The Rise of Compexity in Nature* (Cambridge, Mass.: Harvard University Press, 2001).

Panspermia

Keay Davidson, *Carl Sagan: A Life* (New York: Wiley, 1999); Simon Conway Morris, review of *Cradle of Life*, by J. William Schopf, *New York Times Book Review*, Aug. 15, 1999: 11; Iris Fry, *The Emergence of Life on Earth* (New Brunswick, N.J.: Rutgers University Press, 2000); F. H. C. Crick and L. E. Orgel, "Directed Panspermia," *Icarus* 19 (July 1973): 341–346; Francis Crick, *Life Itself: Its Origins and Nature* (New York: Simon & Schuster, 1981); Gino Segrè, *A Matter of Degrees* (New York: Viking, 2002).

Radio Astronomy

E. W. Barnes, "The Evolution of the Universe," *Nature* 128 (Oct. 24, 1931): 719–722; Woodruff T. Sullivan III, ed., *Classics in Radio Astronomy* (Boston: Reidel, 1982); Woodruff T. Sullivan III, ed., *The Early Years of Radio Astronomy* (Cambridge: Cambridge University Press, 1984); Shirley Thomas, *Men of Space*, vol. 6 (Philadelphia: Chilton Co., 1963); David W. Swift, *SETI Pioneers* (Tucson: University of Arizona Press, 1990); Steven J. Dick, *The Biological Universe* (New York: Cambridge University Press, 1996); Giuseppe Cocconi and Philip Morrison, "Searching for Interstellar Communication," *Nature* 184 (Sept. 19, 1959): 844–846; Frank D. Drake, "Project Ozma," *Physics Today* 14 (Apr. 1961): 40–46; Harlow Shapley, *Of Stars and Men* (Boston: Beacon Press, 1958).

Drake's Equation

Steven J. Dick, *The Biological Universe* (New York: Cambridge University Press, 1996); David E. Fisher and Marshall Jon Fisher, *Strangers in the Night: A Brief History of Life on Other Worlds* (Washington, D.C.: Counterpoint, 1998); Graham Farmelo, ed., *It Must Be Beautiful: Great Equations of Modern Science* (London: Granta, 2002); Walter Sullivan, *We Are Not Alone* (New York: McGraw-Hill, 1964); John L. Casti, *Paradigms Lost: Images of Man in the Mirror of Science* (New York: Morrow, 1989); John C. Lilly, *Man and Dolphin* (New York: Doubleday, 1961); Martin John Wells, *Civilization and the Limpet* (Reading, Mass.: Perseus Books, 1998); Craig B. Stanford, *The Hunting Apes* (Princeton: Princeton University Press, 1999); Peter L. Tyack, "Dolphins Whistle a Signature Tune," *Science* 289 (Aug. 25, 2002): 1310–1311; Mark Derrk, "Brainy Dolphins Pass the Human 'Mirror' Test," *New York Times*, May 1, 2001: 3; Alan Stuart and J. Keith Ord, *Kendall's Advanced*

Theory of Statistics, 6th ed., vol. 1 (New York: E. Arnold, 1994); Frank D. Drake, *Intelligent Life in Space* (New York: Macmillan, 1962); Alastair G. W. Cameron, ed., *Interstellar Communication* (New York: W. A. Benjamin, 1963); Christopher F. Chyba, "Life beyond Mars," *Nature* 382 (Aug. 15, 1996): 576–577; Keay Davidson, *Carl Sagan: A Life* (New York: Wiley, 1999); Steven J. Dick, "SETI," in *History of Astronomy: An Encyclopedia*, ed. John Lankford, 457–459 (New York: Garland, 1997); Amir D. Aczel, *Probability 1* (New York: Harcourt Brace, 1998); Frank D. Drake and Dava Sobel, *Is Anyone Out There? The Scientific Search for Extraterrestrial Intelligence* (New York: Delacorte, 1992); Melvin Calvin, "Talking to Life on Other Worlds," *Science Digest* 53 (Jan. 1963): 14–19, 88, and "What a Man from Space Looks Like," *Science Digest* 53 (Feb. 1963): 88–89; H. W. Nieman and C. Wells Nieman, "What Shall We Say to Mars?" *Scientific American* 122 (Mar. 20, 1920): 298, 312.

Interstellar Visits

Ken Croswell, *Planet Quest: The Epic Discovery of Alien Solar Systems* (New York: Free Press, 1997); Jack J. Lissauer, "How Common Are Habitable Planets?" *Nature* 402 (Dec. 2, 1999): C11–C14; John R. Pierce, "Relativity and Space Travel," *Proceedings of the Institute of Radio Engineers* 47 (June 1959): 1053–1061; Edward Purcell, "Radioastronomy and Communication through Space," in *The Quest for Extraterrestrial Life*, ed. Donald Goldsmith, 188–196 (Mill Valley, Calif.: University Science Books, 1980); Carl Sagan, "Direct Contact among Galactic Civilizations by Relativistic Interstellar Spaceflight," in *The Quest for Extraterrestrial Life*, ed. Donald Goldsmith, 205–213 (Mill Valley, Calif.: University Science Books, 1980); Sebastian von Hoerner, "The General Limits of Space Travel," *Science* 137 (July 6, 1962): 18–23; Walter Sullivan, *We Are Not Alone* (New York: McGraw-Hill, 1966); Thomas Gold, "Cosmic Garbage," *Air Force and Space Digest* 43 (May 1960): 65.

Astroengineering and Supercivilizations

Stephen Webb, *If the Universe Is Teeming with Life . . . Where is Everybody?* New York: Copernicus, 2002; Freeman J. Dyson, "Letters," *Scientific American* 210 (Apr. 1964): 8–10; Eric M. Jones, "Where Is Everybody," *Physics Today* 38 (Aug. 1985): 11–13; Freeman J. Dyson, "Death of a Project," *Science* 149 (July 9, 1965): 141–144, *Disturbing the Universe* (New York: Harper & Row, 1979), "Search for Artificial Stellar Sources of Infrared Radiation," *Science* 131 (June 3, 1960): 1667–1668, and "The Search for Extraterrestrial Technology," in *Perspectives in Modern Physics*, ed. R. E. Marshak, 641–655 (New York: Interscience Publishers, 1966); Carl Sagan and Russell G. Walker, "The Infrared Detectability of Dyson Civilizations," *Astrophysical Journal* 144 (June 1966): 1216–1218; David W. Swift, *SETI Pioneers* (Tucson: University of Arizona Press, 1990); N. S. Kardashev, "Transmission of Information by Extraterrestrial Civilizations," in *The Quest for Extraterrestrial Life,* ed. Donald Goldsmith, 136–139 (Mill Valley, Calif.: University Science Books, 1980); Iosif Shklovsky, *Five Billion Vodka Bottles to the Moon* (New York: Norton, 1991); S. A. Kaplan, ed., *Extraterrestrial Civilizations: Problems in Interstellar Communication* (Jerusalem: Israel Program for Scientific Translations, 1971); Carl Sagan,

"On the Detectivity of Advanced Galactic Civilizations," in *The Quest for Extraterrestrial Life,* ed. Donald Goldsmith, 140–141 (Mill Valley, Calif.: University Science Books, 1980).

CHAPTER 9

The Birth of CETI

Steven J. Dick, *The Biological Universe* (New York: Cambridge University Press, 1996); Steven J. Dick and James E. Strick, *The Living Universe: NASA and the Development of Astrobiology* (New Brunswick, N.J.: Rutgers University Press, 2004); Vladmir Lytkin, Ben Finney, and Liudmila Alepko, "Tsiolkovsky, Russian Cosmism and Extraterrestrial Intelligence," *Quarterly Journal of the Royal Astronomical Society* 36 (Dec. 1995): 369–376; G. M. Tovmasyan, ed., *Extraterrestrial Civilizations* (Jerusalem: Israel Program for Scientific Translations, 1967); Hans Freudenthal, *Lincos: Design of a Language for Cosmic Intercourse* (Amsterdam: North Holland Pub. Co., 1960); Carl Sagan, ed., *Communication with Extraterrestrial Intelligence* (Cambridge, Mass.: MIT Press, 1973); William H. McNeil, "Journey from Common Sense," *University of Chicago Magazine* 62 (May–June 1972): 2–14; Frank D. Drake and Dava Sobel, *Is Anyone Out There? The Scientific Search for Extraterrestrial Intelligence* (New York: Delacorte, 1992); Stephen Webb, *If the Universe Is Teeming with Life . . . Where Is Everybody?* (New York: Copernicus, 2002); William Poundstone, *Carl Sagan: A Life in the Cosmos* (New York: Henry Holt, 1999); Keay Davidson, *Carl Sagan: A Life* (New York: Wiley, 1999).

From CETI to SETI

Steven J. Dick, "The Search for Extraterrestrial Intelligence and the NASA High Resolution Microwave Survey (HRMS): Historical Perspectives," *Space Science Reviews* 64, nos. 1/2 (1993): 93–139; Bernard M. Oliver and John Billingham, *Project Cyclops* (Stanford: Stanford University Press, 1972); Philip Morrison, John Billingham, and John Wolfe, eds., *The Search for Extraterrestrial Intelligence, SETI* (Washington, D.C.: NASA, 1977); the Staff at the National Astronomy and Ionosphere Center, "The Arecibo Message of November 1974," in *The Quest for Extraterrestrial Life,* ed. Donald Goldsmith, 293–296 (Mill Valley, Calif.: University Science Books, 1980); Frank D. Drake and Dava Sobel, *Is Anyone Out There? The Scientific Search for Extraterrestrial Intelligence* (New York: Delacorte, 1992); David E. Fisher and Marshall Jon Fisher, *Strangers in the Night: A Brief History of Life on Other Worlds* (Washington, D.C.: Counterpoint, 1998).

Where Are They?

Steven J. Dick, *The Biological Universe* (New York: Cambridge University Press, 1996); Stephen Webb, *If the Universe Is Teeming with Life . . . Where Is Everybody?* (New York: Copernicus, 2002); Michael H. Hart and Ben Zuckerman, eds., *Extraterrestrials, Where Are They?* (New York: Cambridge University Press, 1995); Frank J. Tipler, "Extraterrestrial Beings Do Not Exist," *Quarterly Journal of the Royal Astronomical Society* 21, no. 3 (1981):

267–281; John von Neumann, *Theory of Self-Reproducing Automata* (Urbana: University of Illinois Press, 1966).

The Immortals

Frank D. Drake, "On Hands and Knees in Search of Elysium," *Technology Review* 78 (June 1976): 22–29; Frank D. Drake and Dava Sobel, *Is Anyone Out There? The Scientific Search for Extraterrestrial Intelligence* (New York: Delacorte, 1992); David W. Swift, *SETI Pioneers* (Tucson: University of Arizona Press, 1990).

Rallying Cry

Anonymous, "Soviet Reserves Position," *Astronomy* 5 (Jan. 1977): 56; Iosif Shklovsky, *Five Billion Vodka Bottles to the Moon* (New York: Norton, 1991), and "Extraterrestrial Civilizations and Artificial Intelligence," in *Cybernetics Today*, ed. I. M. Makarov, 333–355 (Moscow: Mir Publishers, 1984); Philip Morrison, "The Number N of Advanced Civilizations in Our Galaxy and the Question of Galactic Colonization: An Introduction," in *Strategies for the Search for Life in the Universe,* ed. Michael Papagiannis, 15–18 (Dordrecht: Reidel, 1980); Graham Farmelo, ed., *It Must Be Beautiful: Great Equations of Modern Science* (London: Granta, 2002); Cristiano B. Cosmovici et al., eds., *Astronomical and Biochemical Origins and the Search for Life in the Universe* (Bologna: Editrice Compositori, 1997).

SETI at NASA

Steven J. Dick, *The Biological Universe* (New York: Cambridge University Press, 1996); National Research Council, *Astronomy and Astrophysics for the 1980s*, 2 vols. (Washington, D.C.: National Academy Press, 1983); Woodruff T. Sullivan III, "SETI Conference at Tallinn," *Sky and Telescope* 63 (Apr. 1982): 350–353; Keay Davidson, *Carl Sagan: A Life* (New York: Wiley, 1999); William Poundstone, *Carl Sagan: A Life in the Cosmos* (New York: Henry Holt, 1999); Carl Sagan, *The Dragons of Eden* (New York: Random House, 1977), *Cosmos* (New York: Random House, 1980), "Extraterrestrial Intelligence: An International Petition," *Science* 218 (Oct. 29, 1982): 426, and *Contact* (New York: Simon & Schuster, 1985); Thomas Pierson, "SETI Institute: Summary of Projects in Support of SETI Research," in *Progress in the Search for Extraterrestrial Life*, ASP Conference Series, vol. 74, ed. G. Seth Shostak, 443–446 (San Francisco: Astronomical Society of the Pacific, 1995); Frank D. Drake and Dava Sobel, *Is Anyone Out There? The Scientific Search for Extraterrestrial Intelligence* (New York: Delacorte, 1992); John N. Wilford, "Astronomers Open New Search for Alien Life," *New York Times*, Oct. 6, 1992, C1; U.S. Congress, House, *National Aeronautics and Space Administration Multi-Year Authorization Act of 1992*, 102nd Cong., 2nd Sess., *Congressional Record*, vol. 138, no. 56; Richard A. Kerr, "SETI Faces Uncertainty on Earth and in the Stars," *Science* 258 (Oct. 2, 1992): 27; Jill Tarter and M. Klein, "HRMS: Where We've Been, and Where We're Going," in *Progress in the Search for Extraterrestrial Life*, ASP Conference Series, vol. 74, ed. G. Seth Shostak, 457–470 (San Francisco: Astronomical Society of the Pacific, 1995; John N. Wilford, "Astronomers

Search for Life Beyond Earth," *New York Times*, Oct. 13, 1992, C1, and "Etiquette for Handling Otherworldly Calls," *New York Times*, Oct. 13, 1992, C8; Ernst Mayr, "The Search for Extraterrestrial Intelligence," *Science* 259 (Mar. 12, 1993): 1522–1523; U.S. Congress, Senate, The Search for Extraterrestrial Intelligence Program, *Congressional Record*, vol. 139, no. 123; George Johnson, "E.T. Don't Call Us, We'll Call You Someday," *New York Times*, Oct. 10, 1993, IV, 2; Lionel Van Deerlin, "Tuned into Space, NASA Fails to Decode Message from Washington," *San Diego Union-Tribune*, Oct. 5, 1993, B5; John N. Wilford, "Ear to the Universe Is Plugged by Budget Cutters," *New York Times*, October 7, 1993, B12, and "Gifts Keep Alive Search for Other Life in the Universe," *New York Times*, Jan. 25, 1994, C5.

SETI Perseveres

John D. Boudreau, "Cosmic Endeavor or Black Hole? High-Tech Bigwigs Invest in Search for Alien Signs," *Washington Post*, Feb. 15, 1994, E1; Henry Fountain, "Scouring for Space Dust," *New York Times*, Jan. 26, 1999, F5; G. Seth Shostak, *Sharing the Universe* (Berkeley: Berkeley Hills Books, 1998); David Lamb, *The Search for Extraterrestrial Intelligence* (New York: Routledge, 2001); Marina Benjamin, *Rocket Dreams* (New York: Free Press, 2003); Reed Karaim, "Use Your PC to Search for E.T.," *USA Weekend*, Apr. 25, 1999, 4; John D. Biersdorfer, "The Secret Life of the Home Computer," *New York Times*, June 8, 2000, G1.

Pond Scum

Joshua Lederberg, "Exobiology: Approaches to Life beyond the Earth." *Science* 132 (Aug. 12, 1960): 398; Robert Irion, "The Science of Astrobiology Takes Shape," *Science* 288 (Apr. 28, 2000): 603–605; Henry Gee, ed., "Nature Insight: Astrobiology," *Nature* 409 (Feb. 22, 2000): 1079–1122; A. J. S. Rayl, "Update on Astrobiology," *Scientist* 16 (Apr. 15, 2002): 24–25; Donald Savage and John Buck, "Astrobiology Institute Announces New Teams," http://www.seti-inst.edu; Dennis Overbye, "Search for Life Out There Gains Respect, Bit by Bit," *New York Times*, July 8, 2003, F1.

CHAPTER 10

Universal Science?

A. S. Eddington, *Space, Time, and Gravitation* (Cambridge: Cambridge University Press, 1921); David Mulroy, ed. and trans., *Early Greek Lyric Poetry* (Ann Arbor: University of Michigan Press, 1992); David Hume, *The Natural History of Religion and Dialogues Concerning Religion* (Oxford: Oxford University Press, 1976); Edward Purcell, "Radioastronomy and Communication through Space," in *The Quest for Extraterrestrial Life,* ed. Donald Goldsmith, 188–196 (Mill Valley, Calif.: University Science Books, 1980); Richard Q. Elvee, ed., *The End of Science?* (Lanham, Md.: University Press of America, 1992); Ian Hacking, *The Social Construction of What?* (Cambridge, Mass.: Harvard University Press, 1999); Barry Allen, "What It's All About," *Science* 285 (July 9, 1999): 205–206; Steven

Weinberg, "Illustrative Prediction," *Science* 285 (Aug. 13, 1999): 1013; John Horgan, "What If They Don't Have Radios?" *Scientific American* 268 (Feb. 3, 1993): 20; Nicholas Rescher, *Scientific Realism: A Critical Appraisal* (Dordrecht: Reidel, 1987).

The Evolutionists on SETI

Steven J. Dick and James E. Strick, *The Living Universe: NASA and the Development of Astrobiology* (New Brunswick, N.J.: Rutgers University Press, 2004); Theodosius Dobzhansky, "Darwinian Evolution and the Problem of Extraterrestrial Life," *Perspectives in Biology and Medicine* 15 (winter 1972): 157–175; Joshua Lederberg, "Signs of Life," *Nature* 207 (July, 3, 1965): 9–13; George Gaylord Simpson, "The Nonprevalence of Humanoids," *Science* 143 (Feb. 21, 1964): 769–775; Ernst Mayr, "The Probability of Extraterrestrial Intelligent Life," in *Extraterrestrials: Science and Alien Intelligence*, ed. Edward Regis, Jr., 23–30 (New York: Cambridge University Press, 1985), "The Search for Intelligence," *Science* 259 (Mar. 12, 1993): 1522–1523, "Does It Pay to Acquire High Intelligence?" *Perspectives in Biology and Medicine* 37 (spring 1994): 337–338, and "The Search for Extraterrestrial Intelligence," *Planetary Report* 16 (May–Apr. 1996): 4–7; Joel Achenbach, *Captured by Aliens* (New York: Simon & Schuster, 1999); Steven Pinker, *How the Mind Works* (New York: Norton, 1997); David M. Raup, "ETI without Intelligence," in *Extraterrestrials: Science and Alien Intelligence*, ed. Edward Regis, Jr., 31–42 (New York: Cambridge University Press, 1985), and "Nonconscious Intelligence in the Universe," *Acta Astronautica* 26 (Mar./Apr. 1992): 257–261; Carl Sagan, *Cosmos* (New York: Random House, 1980); Stephen Jay Gould, "SETI and the Wisdom of Casey Stengel," in Stephen Jay Gould, *The Flamingo's Smile: Reflections in Natural History*, 403–413 (New York: Norton, 1985); Simon Conway Morris, *The Crucible of Creation: The Burgess Shale and the Rise of Animals* (Oxford: Oxford University Press, 1998), and *Life's Solution: Inevitable Humans in a Lonely Universe* (Cambridge: Cambridge University Press, 2003); Simon Conway Morris and Stephen Jay Gould, "Showdown on the Burgess Shale," *Natural History* 107 (Dec.–Jan. 1998–1999): 48–55; Elliot Sober, review of *Life's Solution*, by Simon Conway Morris, in *New York Times Book Review*, November 30, 2003, 18; Douglas H. Erwin, review of *Life's Solution*, by Simon Conway Morris, in *Science* 302 (Dec. 5, 2003): 1682–1683; Richard E. Lenski, review of *Life's Solution*, by Simon Conway Morris, in *Nature* 425 (Oct. 23, 2003): 767–768.

Progress

Thomas Gold, "The Deep Hot Biosphere," *Proceedings of the National Academy of Sciences* 89 (July 1992): 6045–6049; Stephen Webb, *If the Universe Is Teeming with Life . . . Where Is Everybody?* (New York: Copernicus Books, 2002); Stephen Jay Gould, *Full House: The Spread of Excellence from Plato to Darwin* (New York: Harmony Books, 1996); Carl Sagan, *The Cosmic Connection* (New York: Doubleday, 1973); Robert Nisbet, *History of the Idea of Progress* (New York: Basic Books, 1980); Arnold Burgen, Peter McLaughlin, and Jürgen Mittelstrass, eds., *The Idea of Progress* (New York: W. de Gruyter, 1997); Gabriel A. Almond, Marvin Chodorow, and Roy Harvey Pearce, eds., *Progress and Its*

Discontents (Berkeley: University of California Press, 1982); Leo Marx and Bruce Mazlish, eds., *Progress: Fact or Illusion?* (Ann Arbor: University of Michigan Press, 1996); Jeffrey C. Alexander and Piotr Sztompka, eds., *Rethinking Progress* (Boston: Unwin Hyman, 1990).

Technology

Carl Sagan, ed., *Communication with Extraterrestrial Intelligence (CETI)* (Cambridge, Mass.: MIT Press, 1973); Alastair G. W. Cameron, ed., *Interstellar Communication* (New York: W. A. Benjamin, 1963); John A. Ball, "The Zoo Hypothesis," *Icarus* 19 (July 1973): 347–349; B. M. Oliver and J. Billingham, *Project Cyclops* (Moffett Field, Calif.: NASA, 1971); Frank D. Drake and Dava Sobel, *Is Anyone Out There?: The Scientific Search for Extraterrestrial Intelligence* (New York: Delacorte, 1992); George Basalla, *The Evolution of Technology* (Cambridge: Cambridge University Press, 1988); Michael Ruse, *Darwinian Paradigm* (New York: Routledge, 1989); George Basalla, "Energy and Civilization," in *Science, Technology and the Human Prospect,* ed. Chauncy Starr and Philip C. Ritterbush, 39–52 (New York: Pergamon Press, 1980), and "Some Persistent Energy Myths," in *Energy and Transport,* ed. George H. Daniels and Mark H. Rose, 27–38 (Beverly Hills: Sage Publications, 1982); Carl Sagan, *The Cosmic Connection* (New York: Doubleday, 1973).

Civilization

Sheldon Rothblatt, *Tradition and Change in English Liberal Education* (London: Faber and Faber, 1976); Bruce Mazlish, *Civilization and Its Contents* (Stanford, Calif.: Stanford University Press, 2004); Norman Yofee, *Myths of Ancient States: Evolution of the Earliest Cities, States, and Civilizations* (Cambridge: Cambridge University Press, 2005); Carl Sagan, *Cosmos* (New York: Random House, 1980); I. S. Shklovskii and Carl Sagan, *Intelligent Life in the Universe,* trans. Paula Fern (New York: Dell, 1966); Norman Yofee and George L. Cowgill, eds., *The Collapse of Ancient States and Civilizations* (Tucson: University of Arizona Press, 1988); Joseph A. Tainter, *The Collapse of Complex Societies* (Cambridge: Cambridge University Press, 1988); Harvey Weiss and Raymond S. Bradley, "What Drives Societal Collapse?" *Science* 291 (Jan. 26, 2001): 609–610; Rose Macaulay, *Pleasure of Ruins* (London: Thames and Hudson, 1964); Charles A. Brady, *The Burroughs Encyclopedia* (Jefferson, N.C.: McFarland, 1966); Henry S. F. Cooper, Jr., *The Search for Life on Mars* (New York: Holt, Rinehart, and Winston, 1980); Richard B. Lee and Richard Daly, eds., *The Cambridge Encyclopedia of Hunters and Gatherers* (Cambridge: Cambridge University Press, 1999); Roger Lewin, *Complexity: Life at the Edge of Chaos* (New York: Macmillan, 1992); Michael Shermer, "Why Hasn't E.T. Called?: The Lifetime of Civilizations in the Drake Equation for Estimating Extraterrestrial Intelligence Is Greatly Exaggerated," *Scientific American* 287 (Aug. 2002): 33; Jack Cohen and Ian Stewart, *Evolving the Alien* (London: Ebury, 2002); Peter Corning, *Nature's Magic: Synergy in Evolution and the Fate of Humankind* (Cambridge: Cambridge University Press, 2003).

CHAPTER 11

Robert Plank, *The Emotional Significance of Imaginary Beings* (Springfield, Ill.: Charles C. Thomas, 1968); Carl Sagan, "Unidentified Flying Object," in *The Encyclopedia Americana*, vol. 27: 367–369 (Danbury, Conn.: Grolier, 1982); Cynthia Ozick, "If You Can Read This, You Are Too Far Out," *Esquire* 79 (Jan. 1973): 74–78; Keay Davidson, *Carl Sagan: A Life* (New York: Wiley, 1999); Stewart E. Guthrie, *Faces in the Clouds* (New York: Oxford University Press, 1993); *Dictionary of the History of Ideas*, s.v. "Anthropomorphism in Science," by Joseph Agassi; Carl Sagan, *The Cosmic Connection* (New York: Doubleday, 1973); Anthony Weston, "Radio Astronomy as Epistemology," *The Monist* 71 (Jan. 1988): 88–100; Stephen Hawking, *The Universe in a Nutshell* (New York: Bantam Books, 2001).

INDEX